A GUIDE TO VHDL
Second Edition

A GUIDE TO VHDL
Second Edition

Stanley Mazor

and

Patricia Langstraat

Synopsys, Inc.

Kluwer Academic Publishers
Boston/Dordrecht/London

Distributors for North America:
Kluwer Academic Publishers
101 Philip Drive
Assinippi Park
Norwell, Massachusetts 02061 USA

Distributors for all other countries:
Kluwer Academic Publishers Group
Distribution Centre
Post Office Box 322
3300 AH Dordrecht, THE NETHERLANDS

Library of Congress Cataloging-in-Publication Data

Mazor, Stanley, 1941 -
    A guide to VHDL / Stanley Mazor and Patricia Langstraat. -- 2nd
ed.
        p. cm.
    Includes bibliographical references (p.   ) and index.
    ISBN 0-7923-9387-2 (alk. paper)
    1. VHDL (Computer hardware description language) I. Langstraat,
Patricia, 1945 -   . II. Title.
    TK7885.7.M39   1993
    621.39'2--dc20   93-23132
                                                                    CIP

Contents

Preface

A Guide to VHDL is intended for the working engineer who needs to develop, document, simulate, and synthesize a design using the VHDL language. It is for system and chip designers who are working with VHDL CAD tools, and who have some experience programming in Fortran, Pascal, or C and have used a logic simulator.

This book includes a number of paper exercises and computer lab experiments. If a compiler/simulator is available to the reader, then the lab exercises included in the chapters can be run to reinforce the learning experience. For practical purposes, this book keeps simulator-specific text to a minimum, but does use the Synopsys VHDL Simulator command language in a few cases.

The book can be used as a primer, since its contents are appropriate for an introductory course in VHDL. We have taught the VHDL language with approximately 1 hour per chapter (e.g. about 10 hours of lecture) and 15 hours of simulator exercises.

When the Department of Defense implemented their VHSIC project (Very High Speed Integrated Circuit), the DOD also funded the VHDL project in 1983 - the VHSIC Hardware Description Language. In 1992 the DOD embarked on a program called "Seven Thrusts" in which VHDL is identified as a key technology.

Although hardware-description languages had been around for several decades, they were overlooked during the last 10 to 15 years because of the popularity of schematic-capture systems. Numerous books describe the motivations and background of the language committee, which evolved the language over several years.

The official adoption of VHDL by the IEEE (Standard 1076) in December, 1987 was a significant event in making VHDL an international and recognized standard. The VHDL language was updated in 1992 with some minor improvements. In most cases, the language is upward compatible. Although this book is based primarily on the VHDL 1987 standard, we have indicated the significant changes in the 1992 language to assist you in writing upwardly compatible code.

Primary Benefits of Using VHDL

- Makes the design specification more technology-independent:

    - Uses multiple vendors

    - Avoids part obsolescense

    - Facilitates updating design

    - Documents in a standard way

- Automates low-level design details:

    - Reduces design time

    - Gets to market quicker

    - Reduces design cost

    - Eliminates low-level errors

- Improves design quality:

    - Explores alternatives

    - Verifies functions at a higher level

    - Verifies that implementation matches function

    - Promotes design component re-use and sharing

About This Book

This book contains the following chapters:

Introduction

Provides an overview of the VHDL hardware-description language and how to use VHDL for circuit design.

Chapter 1 VHDL Designs

Introduces all of the basic building blocks of VHDL: the library, package, entity, architecture, and configuration.

Chapter 2 Primitive Elements, 1 + 1 ≠ 2.0

Describes the primitive elements of VHDL, which include scalars, identifiers, and expressions.

Chapter 3 Sequential Statements

Defines the process, the sequential statements that may appear in a process, and the subprogram. It defines the PROCESS statement, Variable Assignment statement, IF statement, CASE statement, LOOP statement, WAIT statement, NEXT statement, EXIT statement, subprograms, and the ASSERT statement.

Chapter 4 Advanced Types

Introduces more advanced, extended data types such as enumerated types that allow for identifying specified values for a type and for subtypes, which are variations of existing types. There are composite types that include arrays and records, and predefined data types (text and lines) that facilitate input and output operations.

Chapter 5 Signals, Signal Assignments, & Concurrent Statements

Discusses the use of signals for component interconnection and process communication. It defines structural netlisting, process communication, signal declarations and delays, and simulation cycles.

Chapter 6 Concurrent Statements

Discusses VHDL concurrent statements: concurrent signal assignment, conditional signal assignment, selective signal assignment, concurrent procedure call, and the BLOCK statement.

Chapter 7 Structural VHDL

Describes the structural style of VHDL, which discusses the interconnection of components within an architecture.

Chapter 8 Packages and Libraries

Explores packages and libraries in detail, and defines the deferred constant, component declaration, and USE statement.

Chapter 9 Advanced Topics

Describes more advanced concepts in VHDL: overloading, resolution functions, and symbolic attributes.

Chapter 10 VHDL & Logic Synthesis

Discusses how VHDL is used for logic synthesis and for technology-independent description of a logic chip. A compiler can process VHDL statements, and synthesize logic-gate netlists and schematics for ASIC libraries or synthetic libraries.

Appendix A Reserved Words

Lists the VHDL reserved words.

Appendix B Application Examples

Provides three examples in VHDL.

Appendix C VHDL Syntax

Provides a summary of the VHDL syntax and language components.

Notation Conventions

This manual contains the following notational conventions:

Convention	Description
monospace bold	In examples, shows user input, for example, **read -format db**.
[ ]	Square brackets denote optional parameters, such as *pin1 [pin2,.. pinN]*. This indicates that at least one name must be entered (*pin1*), but others are optional *[pin2,.. pinN]*.
|	A vertical bar indicates a choice among alternatives, such as **low** | **medium** | **high**. This indicates that you can enter one of three possible values for an option: **low**, **medium**, or **high**.
{ }	Braces indicate that a choice of one or more items must be made.

Acknowledgements

This book evolved from two VHDL classes developed by the Synopsys Technical Training Department staff, which has been teaching VHDL since 1989. Many contributions came from Doug Perry of the Synopsys Applications Engineering staff (we highly recommend his book on VHDL). Joe Cowan and Alison Takata also contributed to the examples. Brian Caslis provided several working VHDL examples, which were used in the "Making the Transition to High Level Design" seminar series. Steve Carlson developed several of the synthesis examples. Appendix C was inspired by *VHDL Designer's Reference* by Berge, et al.

The reader might also enjoy Randy Harr's book *Applications of VHDL to Circuit Design*, and the *Synopsys VHDL Reference Manual*.

Thanks also to Stephan Andres, Clive Charlwood, Charles Dancak, Sandra Goldstein, David Hemmendinger, John Hines, Brent Gregory, Larry Groves, Randy Jewell, Steven Levitan, William Rohm, Rick Rudell, Russ Segal, John Siddall, Charles Smith, and Jose Torres.

The book was done with the cooperation of Synopsys management, and we especially thank Harvey Jones, Aart de Geus, and Chi-Foon Chan for their support.

Looking back...

Thanks to Dr. Andy Grove, who has always provided a role model for many Intel employees.

Special thanks to Professor Ed McCluskey at Stanford, who inspired many students to study algorithmic logic minimization.

Also, thanks to Carver Mead, an associate of Stan Mazor's at SCI, whose suggestions guided our work. In addition, his landmark textbook on MOS digital design was a motivation for this book. *(Introduction to VLSI Systems* by Conway and Mead)

Friends who were encouraging include Profs. Glenn Langdon, Dave Patterson, John Wakerly, and G. Markesjo, from the Royal Swedish University (KTH). Also, the staff at the Manoa campus in Oahu, Hawaii.

About the Authors

Stanley Mazor is the Technical Training Manager at Synopsys, Inc. He joined Synopsys in May, 1988 while the synthesis products were in development, and participated in product introduction and customer-support functions. Previously, he was Director of Custom Engineering at Silicon Compiler Systems (SCS) for four years.

Mr. Mazor was at Intel for 15 years, two of which he spent in Brussels as an Applications Engineer supporting European customers. He joined Intel in its first year, 1969. As one of Intel's early designers he worked on the specification of the early Intel microprocessors. He also supervised Intel's microcomputer training development group.

He has taught courses at the University of Santa Clara and Stanford, and has been a guest professor in China, Finland, and Sweden. He has published over 45 articles and papers on the design and application of VLSI, including signal processing, instrumentation, security, and optimization. He was awarded "Best Paper" for his GOMAC contribution on VHSIC insertion in 1986. Mr. Mazor is a senior member of the IEEE and is active in the COMPCON program committee and Asilomar Microcomputer Workshop. Stanley Mazor has been an active computer user, for both data processing and design automation, for 30 years. He has worked on the development of six computers and microcomputers and developed four small medical instruments, using microcontrollers. He worked on the design of more than 12 VLSI chips.

Patricia Langstraat is the Technical Publications Manager at Synopsys, Inc. She joined Synopsys in September, 1989, and is responsible for all Synopsys technical documentation. While at Synopsys, she has worked with Steve Carlson by editing his publications on VHDL and on Methodology Notes.

Previously, Patricia was Publications and Training Development Manager at BiiN, a subsidiary of Intel and Siemens. She was at Intel before that, and worked there for a total of 7 years. While working at Intel in Oregon, Patricia was selected to participate in the Intel-sponsored Oregon Executive MBA Program and acquired her Masters in Business Administration from the University of Oregon. She also has a B.A. in Psychology from the University of California.

Patricia has worked in project management, training, and publications for 20 years.

Introduction

VHDL is a hardware description language used to document an electronic system design.

VHDL consists of several parts organized as follows:

- The actual VHDL Language
- Some additional data type declarations in the Package STANDARD
- Some utility functions in the Package TEXTIO
- A WORK library reserved for your designs
- A STD library containing Package STANDARD and TEXTIO
- A vendor package
- Vendor libraries
- User libraries and packages

User Package	User Library
Vendor Package	Vendor Library
Package TEXTIO	Library STD
Package STANDARD	
VHDL Language	Library WORK

A VHDL description lists a design's components,   interconnections, and documents the system behavior.

Below is a sample VHDL statement:

```
A <= B + C after 5.0 ns;
```

where **A, B,** and **C** are 8-bit busses representing the interconnection of an adder component and three busses, as shown in the block diagram below:

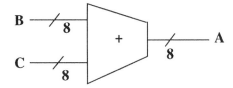

The output **A** of the adder responds **5 ns** after values are asserted on input signal busses **B** or **C**, according to this behavioral description.

This introduction discusses VHDL in the following perspectives:

- Top-Down Design
- Simulation
- Logic Synthesis
- VHDL versus Programming

Top-Down Design

A VHDL description can be written at various levels of abstraction:

- Algorithmic
- Register transfer
- Gate level functional with unit delay
- Gate level with detailed timing

In *top-down* design methodology, a designer represents a system abstractly at first, and later in more detail.

A designer may prefer to delay some design decisions, such as what kind of adder to use. In the example, the **+** is an abstraction of a physical adder. It might be implemented in a design using:

- Ripple carry
- Parallel carry look ahead
- Carry select adders

VHDL provides ways of abstracting a design, or "hiding" implementation details (e.g., the adder). You can evaluate an algorithm in real (floating point) and later implement it in fixed-point integer. Designers often choose integers with 16 bits, but these implementation decisions can be changed during a VHDL design cycle.

In the previous example, addition is specified for 8-bit data. In actuality, the + in this example might be a function call to a user-written function called plus (+) . VHDL calls this technique *operator overloading*. In some previous languages, you would probably think of calling a function by writing `plus (B, C);` the function name followed by the arguments in parentheses. Overloading is another example of information hiding, which is convenient for the reader and allows the design author to describe exactly what kind of operator is wanted (+).

These predefined functions can be grouped together and delivered in a VHDL *package*; vendors often provide packages with their simulator software or synthesis tools. These packages are written in VHDL, and are portable and reusable across all VHDL vendor tools.

Simulation

A design description or model, written in VHDL, is usually run through a VHDL simulator to demonstrate the behavior of the modeled system. Simulating a VHDL design model requires simulated stimulus, a way of observing the model during simulation, and capturing the results of simulation for later inspection. A designer usually creates a set of test cases, for inputs and expected results, to verify the function of the design.

During simulation of the VHDL statement on the previous page, you would typically provide a set of functional test vectors, for example:

At simulated time 0: B = "0000 0010"; the value is 2

C = "0000 0010"; the value is 2

The simulator would run to cover at least 5 ns of simulated time, and you could observe the output on the A bus:

A = "0000 0100"; the value is 4

Logic simulator systems need to represent undefined or "unknown" values for circuits with unresolved electrical values or states, which might be the output of this circuit model during the first 4 ns of simulated time. Accordingly, VHDL supports a variety of data types useful to the hardware modeler for both simulation and synthesis, as well as bits, Booleans, and numbers, which are defined in the Package STANDARD.

In the diagram on page *xvi* of this introduction, the signals **A, B,** and **C** show physical busses in a hardware system. During simulation, the adder is modeled on a computer system, where the values on the busses are stored in the computer's memory. Hence **A, B,** and **C** are similar to variables in programming languages that change value over time, and that can be observed while the program is running.

Logic Synthesis

Some parts of VHDL can be used with logic synthesis tools for producing a physical design. In particular, many VLSI gate-array vendors can convert a VHDL design description into a gate level *netlist* from which a customized integrated circuit component can then be built. Hence the application of VHDL is for:

* Documenting a design.

* Simulating the behavior of a design.

* Directly synthesizing logic.

The diagrammed example on page *xvi* is an abstraction of a real adder circuit, which might have varying delays that are dependent upon the data, or on whether an output rises or falls. Although a logic synthesis tool could generate an 8-bit adder circuit, it could not produce an ideal adder with exactly 5.0 ns delay for all cases. Only the addition function (not the delay specification) of the statement in the example could be synthesized into a physical device.

Programming

Designing in VHDL is like programming in many ways. Compiling and running a VHDL design is similar to compiling and running other programming languages. First, the source *design units* are read by the compiler, error messages are given, and an object module is produced and placed in a special VHDL *library*. Subsequently, a simulation run is made, in which the object units from a library are selected (configured) and loaded into a simulator. A set of test cases are run, either in batch or interactive mode. The main difference is that a VHDL design is always running in simulated time, and events occur in successive time steps. In Chapters 1-6, the language is presented much as if it were a programming language; that is, from a syntax and behavioral example viewpoint. Keep in mind that the code will be executed in a simulator.

Although VHDL is similar to a programming language, there are several differences: the notion of delay and the simulation environment. VHDL also has concurrency and component netlisting, which are not common in programming languages.

Just as a real hardware adder runs all the time, a design language needs a way of specifying concurrent behavior (for example, to represent the adder running in parallel with other simulated hardware elements). VHDL supports concurrency using the concept of *concurrent statements* running in *simulated time*; while concurrency is getting more common, simulated time is a feature found only in simulation languages. Also, there are *sequential statements* in VHDL to describe algorithmic behavior.

Some programming languages provide for design hierarchy by having one main program and separately compiled subprograms. In VHDL, design hierarchy is accomplished by separately compiling *components* that are *instanced* in a higher-level component. The linking process is done via the compiler, or by a simulator using the VHDL library mechanism.

Some software systems utilize version-control systems to generate different versions of loadable programs. VHDL has a *configuration* capability for generating design variations.

Those familiar with programming languages will find VHDL a modern "Ada-like" language. Yet, it has some unique characteristics for the hardware designer.

Summary

The VHDL language can be used to document the interconnection of components and the behavior of an electronic system. A VHDL design description can be input into a simulator to run with test cases. A VHDL design can be input into a logic synthesis tool to produce tooling. A design can be described with several levels of abstraction and with some details hidden to make it easier to read and understand. A designer can design in VHDL, top down with successive refinements, specifying more details of how the design is built.

1 VHDL Designs

A VHDL design consists of several separate *design units*, each of which is compiled and saved in a *library*. The four compilable source design units are: *entity, architecture, configuration*, and *package*.

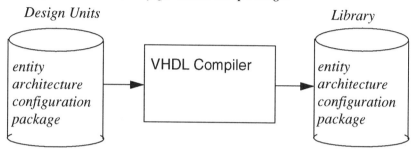

A design's interface signals are described in an *entity*. The design's behavior is specified in an *architecture*. A *configuration* selects a variation of a design from a *library*. For convenience, certain frequently used specifications can be stored together in a *package*. An example is shown in *Figure 1-1*.

Typically, a designer's architecture uses previously compiled components, such as gates, from an ASIC (Application Specific Integrated Circuit) vendor library. Once compiled, a user's design also becomes a *component* in a library that may be used in other designs. Additionally, vendors provide packages of functions, such as square root, written in VHDL to assist the designer. These compiled packages are also stored in a library for your use.

Separating the entity (I/O interface of a design) from its actual architecture implementation simplifies experimenting with alternative implementations. Configurations provide flexibility by saving variations of a design (for example, two versions of a CPU, each with a different multiplier). A configuration is a named and compiled unit, stored in the library, that allows you to do minor substitutions of design parts without having to copy and compile the entire design.

Libraries contain compiled entities, architectures, packages, and configurations. They store standard VHDL packages, user designs, and ASIC-vendor supplied components.

The following is an example of some design units in a text file: **mine.vhd**

```
package my_defs is                                    Package
    constant unit_delay: time := 1 ns;   --room temperature
end my_defs;


entity COMPARE is                                     Entity
    port (a, b: in bit; c: out bit);
end COMPARE;


architecture flow of COMPARE is                       Architecture
    begin
    c <= NOT (a XOR b) after work.my_defs.unit_delay;
end flow;
```

Figure 1.1

There are three design units in a text file **mine.vhd**. After analysis, there are four compiled units in the default library named WORK:

- Package **my_defs** - Provides a shareable constant **unit_delay**.

- Entity **COMPARE** - Names the design and signal ports **a,b,c**.

- Architecture **flow** of **COMPARE** - Provides details of the design.

- A default configuration of **COMPARE** - Designates **flow** as the latest compiled version of the architecture of the entity **COMPARE**.

Each design unit could be in a separate file and could be compiled separately, but the order of compilation must be as shown in ***Figure 1-1***, eg. Package, Entity, Architecture. The package **my_defs** can also be used in other designs, or changed later. The design entity **COMPARE** can now be accessed from the library WORK for simulation, or used as a precompiled component in another design.

When using the component **COMPARE** in a new design, two input values of type **bit** are required at pins **a** and **b**; **1 ns** later a '1' or '0' appears at output pin **c**. This delay is specified in **flow** to be **after work.my_defs.unit_delay** (constant **1 ns** in the package **my_defs** from the library **work**).

This chapter defines the basic VHDL building blocks in the following sections:

- Library

- Package

- Entity

- Architecture

- Configuration

1.1 Library

The results of a VHDL compilation are kept inside of a library for subsequent simulation, or for use as a component in other designs. A library can contain:

- A package - shared declarations

- An entity - shared designs

- An architecture - shared design implementations

- A configuration - shared design versions

The benefit of using a library is it promotes the sharing of previously compiled designs and the source design units need not be disclosed to all users. The two built-in libraries are WORK and STD, but you can have other libraries.

Note: VHDL does not support hierarchical libraries (i.e., you can have as many libraries as you want, but cannot nest them).

Compilation

When you compile (analyze) your VHDL source design units, your compiled design is put into the library WORK unless you direct it to another library.

Examples:

```
vhdlan mine.vhd                    into library WORK, or

vhdlan -w freds_lib mine.vhd       into freds_lib
```

These statements vary depending upon the vendor's implementation.

Simulation

To run a simulation of a compiled entity called COMPARE from a library:

```
vhdlsim freds_lib.COMPARE          from freds_lib

vhdlsim COMPARE                    from library WORK
```

The configuration unit of COMPARE selects the version to run.

Using Library Units

To open a library to access a compiled entity (component) as part of a new VHDL design, you first need to declare the library name. The syntax is:

```
library  logical_names;
```

No library statement is required for the libraries WORK or STD. The default is:

```
library WORK, STD;
```

Component designs compiled into the library can now be used; packages in the library can be accessed. In VHDL, your default library is called WORK. The library STD is reserved for **stand**ard built-in packages. Additionally, you can have other named libraries where you have your own compiled designs. These resource libraries need to be declared, and typically have ASIC vendor components, some special data types and operators for simulation, design elements, etc.

You can access compiled design units from a VHDL library with up to three levels of names. VHDL calls `name.name` *selected names*:

        `library_name.package_name.item_name`

Example: `work.my_defs.unit_delay;`

or

        `library_name.item_name`

Example: `freds_lib.COMPARE`

or

        `item_name`   the library name is assumed

Example: `COMPARE`

As shown in *Figure 1-1*, the package `my_defs` is compiled prior to being referenced in the architecture `flow`. The reference `work.my_defs.unit_delay` identifies the library `work` and the package name `my_defs`, as the context to locate `unit_delay`, a 1 ns constant. A USE clause (see Section 8.5) allows simpler access to items in a library, and allows the library name to be assumed.

Units in a library must have unique names; all design entity names and package names are unique within a library. Architecture names need to be unique to a particular design entity. Use selected names to get a particular element.

Library in File System

To locate a VHDL library in a file system requires some mapping commands outside of the VHDL language, eg. in a start-up file. One vendor's example is:

        `library_name : path/directory_name`

Another example is:

        `freds_lib : /sys/user/fred/freds_stuff`

1.2 Package

Another level of hierarchy within a library is a *package* (detailed in Chapter 8). A package collects a group of related declarations together and is compiled separately. Typically, a package is used for:

- constant_declarations (Chapter 2)

- subprogram_declarations (Chapter 3)

- type_declarations (Chapter 4)

You create a package to store common subprograms, data types, constants and compiled design interfaces that you can use in more than one design. This strategy saves coding of commonly used declarations, which saves time and promotes reuse. If revisions are required, in many cases you only need to recompile the package bodies and not the design units that use them, saving time and effort. See *Figure 1-2* for a sharable `constant` declaration `unit_delay` of type `time` declared within `package` `my_defs`. A package can consist of two separate design units: the package header, which identifies all of the names of items; and the optional package body, which gives more details of the named item. Example:

```
    Package my_defs is                     Package Header

    Constant unit_delay: time := 1 ns;

    end my_defs;                           No Package Body
```

Figure 1-2

Use a selected name to access an item from a package; see *Figure 1-1*. Example:

```
    ... after    work.my_defs.unit_delay;    Package Used
```

All vendors provide a package named STANDARD in the predefined library named STD. The Package STANDARD provides useful data types, such as bit, boolean, bit_vector, and time. As shown in *Figure 1-2*, the `constant` named `unit_delay` is of type `time` -- a pre-defined type in the Package STANDARD and does not require a selected name.

Additionally, vendors provide packages of design pieces and utility routines to assist your design work. An ASIC vendor provides VHDL descriptions of 100-gate components for a CMOS gate-array library. A CAD company provides some utility functions, such as square root, exponential functions, and random number generators.

A USE clause (see Section 8.5) allows simpler access to items in a package, without using selected names.

1.3 Entity

The design *entity* defines a new component name, its input/output connections, and related declarations. The entity is the I/O interface to a component design. VHDL separates the external interface to a design from the details of architectural implementation. The entity describes the direction and type of signal connections; in Section 1.4, you can see that the architecture describes the behavior of a component. After an entity is compiled into a library, it can be simulated or used as a *component* in another design. See Section 1.4. An entity must have a unique name within a library. If a component has signal ports, they are declared in an entity declaration (see Section 5.5). Subsequent chapters describe the declarations of types (Chapter 4), constants (Chapter 2), signals (Chapter 5), and generics (Chapter 7).

The syntax is:

entity_declaration

entity name **is**	Required
[generics] [ports]	Optional
[declarations{constants, types, signals}]	Optional
[**begin** statements]	Typically not used
end [name]**;**	Required (name is optional)

An entity specifies the external connections of a component.

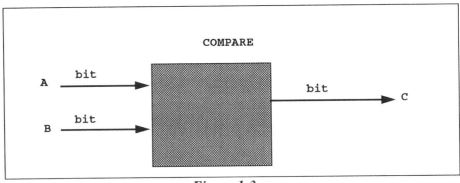

Figure 1-3

In the above diagram, you see a comparator (**COMPARE**) with two signal lines coming in, and one going out. The diagram characterizes the interface to the design, which is shown as a black box. The signal types are all type **bit**, which mandates the usage; the **COMPARE** component only works on **bit** type data.

| Port Names, Directions, and Data Types | ```
entity COMPARE is
  port (A,B: in bit;
           C: out bit);
  end COMPARE;
``` |

Figure 1-4

The above example defines COMPARE as a new component. The reserved word entity is followed by the entity name COMPARE. The reserved word is, is followed by the port declarations, with their names and types. A and B are input signals of type bit; C is an output of type bit. (VHDL is not case sensitive.) Type bit is predefined in Package STANDARD.

In some cases, constraints are needed to specify the number of bits of a port or the range for integer values. You can also specify an initial value. For example:

```
entity COMPARE_digit is
  port (B,A: in integer range 0 to 9 := 0;   range constraint
        C:out boolean);
  end COMPARE_digit;
```

Figure 1-5

This example identifies the input signals B,A as type integer , with a constrained range of 0 to 9 , and the explicit initial value of 0; the output signal is type boolean with a default initial value of FALSE.

A package is used to declare constants or types for use in an entity port.   For example, port A is given an inital value, big_number, from a package:

```
port (A: in integer := work.pckstuff.big_number);
```

NOTE: big_number must be in a previously compiled package pckstuff.

When an entity is compiled into a library, it becomes a component design that can be used in another design. You can use a component without knowing its internal architecture, which saves you from dealing with the complexities of the design. However, you can only use the component with types matching the port specification. Also, you save time because you do not need to recompile the component.

Question: Can COMPARE_digit output C be used as an input to the COMPARE entity described in *Figure 1-5*?

Answer:

Question: Could the constant big_number be declared in the entity declarations?

Answer:
```

1.4 Architecture

An architecture design unit specifies the behavior, interconnections, and components of a previously compiled design entity. The architecture defines the function of a design entity. It specifies the relationships between the inputs and outputs of a design entity that may be expressed in terms of behavior, dataflow, or structure. You must compile the entity design unit before you compile its architecture. If you recompile an entity, you must recompile its associated architectures.

A designer can model a design at several levels of abstraction or with various implementations (see introduction, e.g., carry-look-ahead or ripple-carry adder). An entity may be implemented with more than one architecture. *Figure 1-6* illustrates three alternate architectures of entity **xyz**. The architectures have an identical interface in common, but each needs a unique architecture name. The designer selects a particular architecture of a design entity, for example **xyz** (**b**), during configuration (see Section 1.5). Design speed and size may vary. A design may be described at several levels, (e.g., block functional, behavioral model, or gate level). The advantage of having separately compiled architectures is that one designer can select a particular architecture of an entity, while another designer is updating a different architecture. This feature promotes version control.

VHDL architectures are generally categorized in styles as follows:

- Behavioral - Defines a sequentially described process (Chapter 3).

- Dataflow - Implies a structure and behavior (Chapter 6).

- Structural - Defines interconnections of components (Chapter 7).

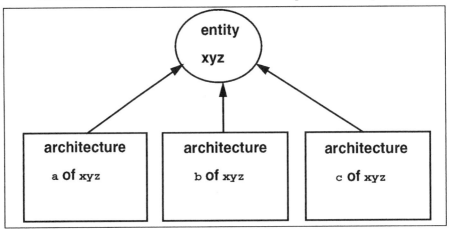

Figure 1-6

A design can use any or all of these design styles. An architecture can also be a mix of these design styles. Select the style according to the level of abstraction or functional detail desired. For example, use a plus sign (+) to describe an adder's behavior, a logical equation to provide details of the adder, or a list of gates to define the adder's structure.

Generally, designs are created hierarchically using previously compiled design entities (see *Figure 1-14*); they can only be combined using the structural style. A structural style architecture ef like a list of components wired together (i.e., a netlist).

The syntax is:

architecture_declaration

      architecture name **of** entity_name **is**

        [declarations]

      begin

        concurrent_statements

      end [name]**;**

The architecture name defines the unique name of this architecture for the entity it refers to. The architecture declarations are items used only in this architecture such as types, subprograms, constants. The statements are the actual design description.

By convention, an architecture may be named as *behavioral, dataflow,* or *structural.* The three styles are shown in the next sections.

Behavioral Style Architecture

The example below illustrates **ARCH1**, which is an architecture (behavior style) of the entity **COMPARE** (see Figure 1-3). It contains a **process** that uses signal assignment statements. If **A** = **B** then **C** gets a '1'; otherwise **C** gets a '0'. This architecture describes a behavior in a "program-like" or algorithmic manner.

```
architecture ARCH1 of COMPARE is
begin
   process (A, B)
   begin
      if (A = B) then
      C <= '1' after 1 ns;
      else
      C <= '0' after 2 ns;
      end if;
   end process;
end ARCH1;
```

Figure 1-7

The list of signals for which the process is waiting (sensitive to) is shown in parentheses after the word **process** (e.g., **A, B**). Processes wait for changes in an incoming signal. In *Figure 1-7*, whenever input signals **A** or **B** change, the process is activated. The output delay of **C** depends upon the **after** clause in the assignment statement (**1 ns** or **2 ns** in the example). Signal **C** changing might trigger some other process execution. (See Chapter 3 for more information about a process, and see Chapter 5 for more information about signals.)

To describe your architecture in a program-like or behavioral style, use the VHDL process. Processes are language-sequential descriptions that the simulator can execute. Parallel operations can be represented with multiple processes, as shown by **m** and **n** in *Figure 1-8*.

Processes are concurrent statements. VHDL processes may run concurrently (i.e., in a simulator at the same *simulated time*). *Figure 1-9* shows two processes running in parallel. The processes communicate with each other; they transfer data with *signals*. A process gets its data from the outside from a signal. Inside, the process works with variables. The variables are local storage that cannot be used to transfer information outside the process. Also inside the process are *sequential statements* that execute in order in a programming-like manner.

The figure illustrates a sending process **m** and a receiving process **n**. The running of a process can depend upon the results of another process. Processes can also be independent.

Architecture C

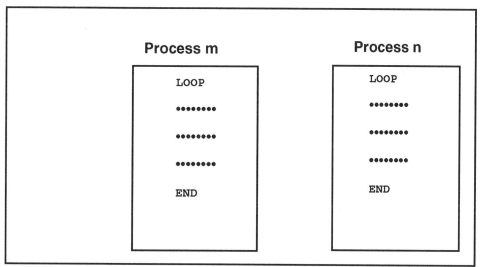

Figure 1-8

Process Model

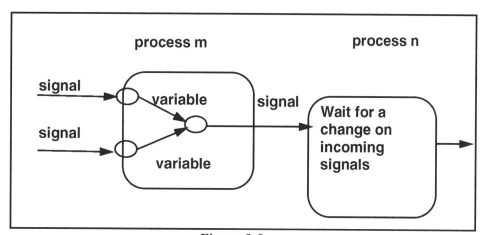

Figure 1-9

In a top-down design style, this behavioral type of description is usually the first step; the designer can focus on the "abstract" behavior of a design, rather than the lower level and structural details of a design. Detailed timing and exact signal communication can also be modeled more abstractly. Later, the designer can choose the precise signal-bus width and coding.

Dataflow Style Architecture

A dataflow architecture models the information or dataflow behavior over time of combinational logic functions such as adders, comparators, decoders, and primitive logic gates. The example below illustrates an architecture (dataflow style) of entity COMPARE. C gets the not of A xor B. The parentheses imply that the xor is computed first to order the flow of data. Note also, the time delay of a literal 1 ns from a change of A or B to C in contrast to the package constant used in *Figure 1-1,* which is more general.

```
architecture ARCH2 of COMPARE is
begin
   C <= not (A xor B) after 1 ns;
end ARCH2;
```

Figure 1-10

The following example defines the entity and architecture, in a dataflow style, of XR2 , an exclusive-or gate. XR2 has an input port X and Y of type BIT and an output port Z of type BIT. There is also a generic delay parameter m, which defaults to 1.0 ns. (See Section 7.5.) The architecture DATAFLOW is a one-line architecture. The output z gets the exclusive-or, X xor Y after m (1 ns).

```
entity XR2 is
   generic (m: time := 1.0 ns);        -- Delay Time
   port (X,Y: in BIT; Z: out BIT);
end XR2;

architecture DATAFLOW of XR2 is
begin
   Z <= X xor Y after m;               -- Generic Delay
end DATAFLOW;
```

Figure 1-11

Once you compile this simple gate into a library, you can use it as a component in another design by referring to the entity name XR2, and providing three port parameters and, optionally, a delay parameter. This action is shown in the next section. The dataflow style is convenient for illustrating asynchronous and concurrent events. Delays typically represent hardware component delays.

See Chapter 6 for more information on concurrent statements used for dataflow architecture.

Structural Style Architecture

Most top-level VHDL designs use the structural style to *instance* and connect previously compiled designs. The following example uses the **XR2** gate from *Figure 1-11*. It implements a structural style architecture of the entity **COMPARE**. The schematic in *Figure 1-12* has two components: an exclusive-or gate **XR2**, which has input connections **A** and **B** and an output **I**. The signal wire **I** from **XR2** connects to the next component **INV**, which is an inverter that has output **c**.

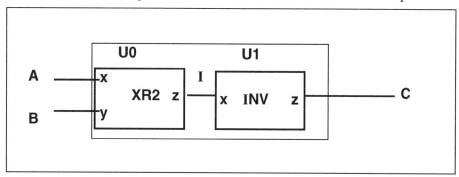

Figure 1-12

```
entity COMPARE is                                 Entity
  port(A,B: in BIT; C: out BIT);
end COMPARE;

architecture STRUCT of COMPARE is
signal I: BIT;                                    Components Declared

component XR2 port (x,y: in BIT; z:out BIT);end component;
component INV port (x: in BIT; z: out BIT);end component;

begin
  U0: XR2 port map (A,B,I);
  U1: INV port map (I,C);                         Components Used
end STRUCT;
```

Figure 1-13

In Figure 1-13 the architecture has an arbitrary name **STRUCT**. Inside of the architecture is a declaration of a local **signal I**, type **BIT**. The **component** declarations shown in the box are required unless these declarations are placed in a package*, saving the designer time for frequently used declarations. Two components are given instance names **U0** and **U1**. The **port map** indicates the signal connections to be used. **A** is wired to **XR2** pin **X**; **B** is wired to **XR2** pin **y**; **I** is wired to **XR2** pin **z**. A configuration can select the library from which the components **XR2** and **INV** are to be found (see *Figure 1-16*).

*Optional in VHDL 92.

Design Hierarchy

VHDL is used to document the structure, components and interconnections of a design. A typical design consists of a hierarchy of design components. The lowest level components may be from a vendor library. Below is a diagram illustrating design hierarchy of previously compiled components. The CPU contains an ALU component. The ALU component contains a COMPARE component. Primitive components, such as XR2 and INV (xor gate and an inverter) from a library, are shown in the innermost block of *Figure 1-14*. These primitive components are at the lowest level in a component-oriented design.

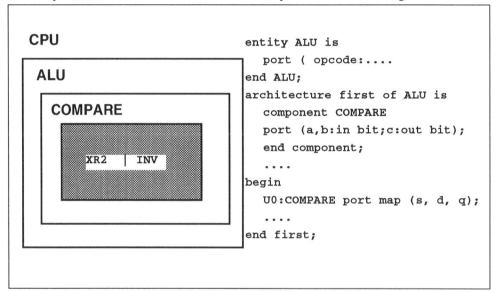

Figure 1-14

The component design of COMPARE was previously shown with the design unit entity and architecture. COMPARE, having been compiled and stored in library WORK, can be used as a component (in this case, in an arithmetic logic unit ALU). The ALU declares that the component COMPARE was originally defined, with entity inputs a and b and output c. (See *Figure 1-13*.) Architecture first takes an *instance* of COMPARE named U0 and has a signal port map to the ports a, b, c using (s, d, q). Having compiled the ALU into a library, you could design a CPU that takes an instance of an ALU component. Each time you compile an entity, it becomes a design that can be used as a component in another design (or a design that can be simulated).

Question: Can you nest entity and architecture declarations in your source design file?
Answer:

1.5 Configuration

The *configuration* assists the designer in experimenting with different variations of a design by selecting particular architectures or components. ***Figure 1-6*** illustrates three alternate architectures of entity **xyz**. A configuration statement selects one of the architectures **a, b,** or **c**; for example **xyz(b)**. A configuration selects a particular architecture of an entity from a library. The syntax is:

configuration_declaration

> **configuration** name **of** entity_name **is**
>
> > [declarations] specification
>
> **end** name**;**

You may have more than one architecture of an entity. These architectures may use different algorithms or different levels of abstraction. If you want to configure a design to use a particular architecture, use the CONFIGURATION statement. A configuration is a named and compiled unit stored in a library. The VHDL source description of a configuration identifies (by name) other units from a library.

For example:

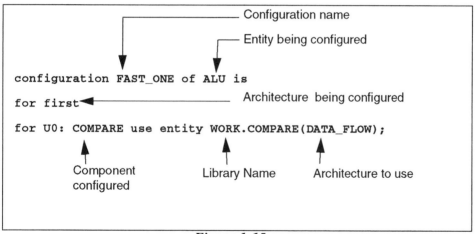

Figure 1-15

Figure 1-15 shows that for entity **ALU** and architecture **first**, you are creating a configuration called **FAST_ONE**. Of the various architectures of **COMPARE**, use the **DATA_FLOW** architecture in the library **WORK** for **U0**. The USE clause identifies a library, entity, and architecture of a component (e.g., **COMPARE**). The ending result is the configuration called **FAST_ONE,** a variation of the design **ALU**.

Configurations permit selection of a particular architecture. When no explicit configuration exists, the latest compiled architecture is used (called the *null* configuration), as shown in ***Figure 1-1***.

When you run a simulation, you select a configuration from a library. Since the configuration design unit, FAST_ONE shown in ***Figure 1-15***, was compiled into library WORK, you can simulate it. For example:

```
vhdlsim FAST_ONE                library WORK is used
```

Configurations permit selection of components in structural designs. In the example below, the design shown in ***Figure 1-13*** is partially configured by selecting a particular XR2 component from library WORK and architecture DATAFLOW for U0.

```
configuration main of COMPARE is

   for STRUCT

      for U0: XR2 use entity WORK.XR2(DATAFLOW);

      end for;

   end for;

end;
```

Figure 1-16

A typical application of a configuration in a hardware design is using a TTL logic family that comes in many flavors, such as low-power or high-speed. You can use a generic gate design and then custom tailor it with a particular library and configuration.

Many companies document changes during production of a system with *ECOs* (Engineering Change Orders). Conceptually, a configuration can also be used to document changes or variations made during the design process or after a design is in production, promoting version control. (Also see Section 7.4 for more information about configurations.)

- **Summary**

1. *Entity declarations, architectures, configurations,* and *packages* are compiled and stored in a library. The default library is called WORK. (Section 1.1)

2. *Packages* contain declarations of *components, subprograms,* and other elements grouped together. The VHDL language is extended through the use of Package Standard, vendor packages, and user-developed packages. (Section 1.2)

3. A complete *design* is an interconnection of compiled components. (Section 1.4)

4. Each component design has an entity specification and an architecture specification. (Sections 1.3 and 1.4)

5. All communication occurs through ports declared in the entity specification with matching signal types, sizes, and directions. (Section 1.3)

6. An architecture may be modeled at the behavioral or structural level. (Section 1.4)

7. A complete design is a configuration, an interconnection of particular architectures. (Section 1.5)

8. *Figure 1-17* illustrates that there are four kinds of design units that can be compiled into a library. A VHDL design is usually comprised of some combination of these design units; at the top level, a configuration specifies which versions of design units are chosen.

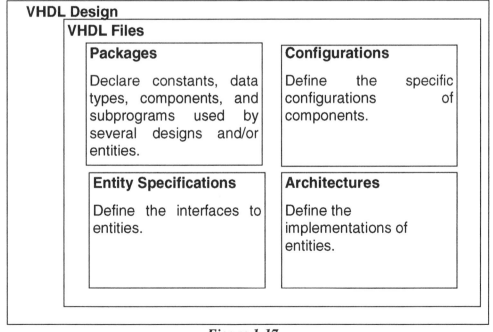

Figure 1-17

- **Answers to Questions**

Question: Can COMPARE_digit output C be used as an input to the COMPARE entity described in *Figure 1-5*?

Answer: No, the output is boolean, COMPARE inputs A and B are of type bit and cannot be wired to boolean.

Question: Could the constant big_number be declared in the entity declarations?

Answer: No, ports must use <u>previously</u> declared types; declarations follow ports in an entity.

Question: Can you nest entity and architecture declarations in your source design file?

Answer: No, you compile separately and refer to previously compiled entities as components.

- **Lab #1**

Text Edit of source file

1. Edit the source file `test.vhd`.

2. Run the analyzer `vhdlan test.vhd`.

3. Inspect your directory to observe the compiled units.

4. Run the simulator `vhdlsim test`

```
entity test is
end;
architecture one of test is
begin
process
  begin
    assert (false)
    report "hello";
    wait for 10 ns;
end process;
end one;
```

2 Primitive Elements
$1 + 1 \neq 2.0$

VHDL is used to document and model electronic systems. You can design top-down by creating a behavioral model and validating it with a simulator. To support this methodology, VHDL provides a number of data *types* and *operators* to support both behavioral and gate-level simulation. You can use real (floating-point) types in the early stages of behavioral simulation, use integer types at the register transfer level, and use logic simulation at the gate level.

The Package STANDARD provides *behavioral* data types and operators. The IEEE 1164 package provides *synthesis* and *simulation* types and operators. The benefit of using this package is that it will likely be an industry standard and supported by multiple CAE vendors. Additionally, users can define their own data types and operators and include them in their own user package. This chapter discusses the standard data types.

The reason that $1 + 1 \neq 2.0$ is that literal constants, such as the integer 1, the real number 1.0, and the bit '1' are not the same in VHDL. In other words, VHDL is a *strongly typed* language, which assists designers in catching errors early in the development cycle (for example, a compiler identifies an error if you mistakenly try to connect a 4-bit and 8-bit bus).

Each type defines a set of legal values and a lowest (leftmost) value. This value is used by simulators as the default initial value (for signals of that type).

This chapter discusses the standard data types available in VHDL vendor packages, as well as the use of constants, signals, and variables, and other named objects in the following major sections:

- Scalars and Array Literals
- Names
- Objects
  - Constants
  - Signals
  - Variables
- Expressions

2.1 Scalars and Array Literals

An array is made up of elements that have the same type. A *scalar*, as opposed to an array, is a single value. A scalar is made up of characters or digits (or it can be named, see Section 2.2), as long as it contains only one element. Predefined types are listed below and are illustrated in this chapter.

Scalar Type	Array Type
character	string
bit	bit_vector
std_logic	std_logic_vector
boolean	
real	
integer	
time	

Constant literals have special formats, depending upon the type of constant and the way it is used. Therefore, you must watch your punctuation of literal constants in your source design file. The compiler's analyzer is very exact and gives you errors for not using the correct data representation. You must use the correct data type for a particular operation. For example, you can add integers but may not be able to add characters. A good way to understand these types is to look at the literal way you represent them in your VHDL design.

Character Literal

A *character literal* defines a value by using a single character enclosed in *single* quotes: '**x**'. Generally VHDL is not case sensitive; however, it does consider case for character literals. For example:

'**a**' is not the same as '**A**'

Character literals can be any alphabetic letter a-z, digit 0 - 9, blank, or special character, such as ', $, @, %. The character '1' differs from integers or real numbers, such as 1 (one) or 1.0.

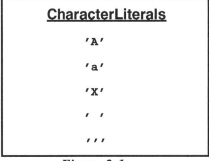

Figure 2-1

The Package STANDARD predefines type character to be any of:

NUL	SOH	STX	ETX	EOT	ENQ	ACK	BEL
BS	HT	LF	VT	FF	CR	SO	SI
DLE	DC1	DC2	DC3	DC4	NAK	SYN	ETB
CAN	EM	SUB	ESC	FSP	GSP	RSP	USP
' '	'!'	'"'	'#'	'$'	'%'	'&'	'''
'('	')'	'*'	'+'	','	'-'	'.'	'/'
'0'	'1'	'2'	'3'	'4'	'5'	'6'	'7'
'8'	'9'	':'	';'	'<'	'='	'>'	'?'
'@'	'A'	'B'	'C'	'D'	'E'	'F'	'G'
'H'	'I'	'J'	'K'	'L'	'M'	'N'	'O'
'P'	'Q'	'R'	'S'	'T'	'U'	'V'	'W'
'X'	'Y'	'Z'	'['	'\'	']'	'^'	'_'
'`'	'a'	'b'	'c'	'd'	'e'	'f'	'g'
'h'	'i'	'j'	'k'	'l'	'm'	'n'	'o'
'p'	'q'	'r'	's'	't'	'u'	'v'	'w'
'x'	'y'	'z'	'{'	'|'	'}'	'~'	DEL

Figure 2-2

Note: NUL is the lowest value of this type, and is the default value.

In some cases, it may be necessary to make the type of a literal explicit by providing the type name. For example:

```
character'('1')
```

String Literal

A character *string* is an array of characters. Literal character strings are enclosed in *double quotes*.

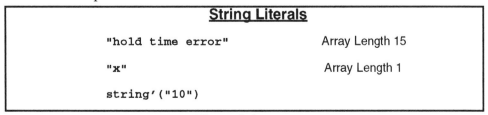

String Literals	
`"hold time error"`	Array Length 15
`"x"`	Array Length 1
`string'("10")`	

Figure 2-3

Bit Literal

The value of a signal in a digital system is often represented by a bit. A *bit literal* represents two discreet values by using the character literals ′0′ , or ′1′. In some cases, it may be necessary to make a bit type literal explicit to distinguish it from a character (remember, VHDL is a strongly typed language). For example:

> `bit′(′1′)`

Bit (<u>bi</u>nary di<u>git</u>) literals differ from integers, such as 1 (one) or 0 (zero). ′0′ is the default value for uninitialized variables of type bit. Bit data is distinct from boolean data, although you may implement conversion functions.

```
     Bit Literals

        ′1′

        ′0′

      BIT′(′1′)
```

Figure 2-4

Bit_Vector

A *bit_vector* literal is an array of bits enclosed in double quotes. For example:

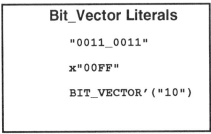

```
     Bit_Vector Literals

       "0011_0011"

       x"00FF"

     BIT_VECTOR′("10")
```

Figure 2-5

Hexadecimal, octal, and binary representations are supported in your design specification. The default is binary. The underscore (_) is allowed in the source for clarity and ignored. Bit literals could be used to describe the value of a bus in a digital system. Most simulators need additional values representing unknowns, high impedence states, or other electrically-related values. Vendors provide extended data types (packages written in VHDL) that make the language more useful for logic-level simulation, such as IEEE 1164.

IEEE Standard Logic

The IEEE standard 1164 defines nine signal strengths within a VHDL package. These nine values are more useful for simulation and synthesis than type bit. The nine values are:

```
               Std_Logic

U          uninitialized
X          forcing an unknown
0          forcing 0
1          forcing 1
Z          high impedance          Three State
W          weak unknown
L          weak 0
H          weak 1
-          don't care
```

Figure 2-6

The character literal should be in upper case letters and in single quote marks. For example:

  'H'   *NOT*   'h'

Uninitialized is the left most, lowest element of this type; so the default initial value for signals (and variables) of this type is U. Behavioral models may produce an unknown value (X) as an output during operations that are ambiguous -- consider a model of an AND gate that might have an output settling time of 2 ns. During the first 2 ns after an input change, the output could be designated as unknown. Don't care (-) is generally a design specification used to give a logic synthesis tool flexibility in developing the logic circuitry.

Std_Logic_Vector

An array of std_logic elements is called a std_logic_vector. Some examples are:

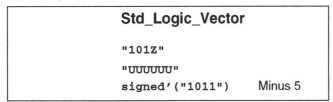

```
          Std_Logic_Vector

     "101Z"
     "UUUUUU"
     signed'("1011")     Minus 5
```

Figure 2-7

Vendors provide logical and arithmetic operators (both signed and unsigned) that make this the most useful data type for logic simulation and synthesis. The example shows how a literal might be declared to be **signed**, if the vendor supports this type. A USE statement is required:

```
Library IEEE;
Use IEEE.STD_LOGIC_1164.all;
Use IEEE.STD_LOGIC_SIGNED.all;
```

Boolean Literal

A *Boolean literal* represents two discreet values, true or false. A Boolean literal is not a bit literal and has no relationship to a bit. A Boolean can be tested in an IF statement. The default value is `false` for uninitialized variables (see example shown in *Figure 1-5*). Relational operators like `=`, `<=`, `>=`, and `/=` produce a Boolean result. For example:

```
true    TRUE    True        are equivalent

false   FALSE   False       are equivalent
```

Boolean Literals

```
true

false

TRUE

FALSE
```

Figure 2-8

A Boolean signal is often used to express the state of an electronic signal or a condition of a bus with a single boolean value. However, Package STANDARD does not predefine an array of boolean values. Refer to Chapter 4 to see how you can create your own array types.

This type is useful in simple signaling, although actual physical hardware systems may use a signal wire and some encoding scheme. Boolean signals are useful as abstractions of actual signals (for example, an electrical high or low value).

Real Literal

During algorithm development and experimentation, floating-point arithmetic can be used as an abstraction for a hardware block, which may subsequently be built with fixed-point arithmetic. Synthesis tools typically do not support either real arithmetic or real literals, but simulators do support type real.

There are an infinite number of real numbers between 0.0 and 1.0. Limitations of the computer allow only about 10 million distinct numbers in this range. Most VHDL systems support at least six decimal digits of precision. *Real literals* represent numbers from -1.0E + 38 to +1.0E +38 (not of a discreet type). A real number may be positive or negative, but must always be written with a decimal point in your VHDL source design . The format is + (optionally) or - followed by `number.number`. For example:

+1.0	*NOT*	'1' or 1
3.5	*NOT*	'3.5'
0.0	*NOT*	0

Real Literals

-1.0

+2.35

36.0

-1.0E + 38

Figure 2-9

Uninitialized variables or signals have a default initial value of the most negative number, such as **-1.0E + 38.**

Integer Literal

Integer literals define discreet values as mathematical numbers do. They are useful for counting, indexing, and controlling loops. In most VHDL implementations, the range is -2,147,483,647 to +2,147,483,647 but you can use a range constraint (see *Figure 1-5*). Be careful not to use a real number (number with a decimal point) where an integer variable is expected because, in a strongly typed language such as VHDL, you must match data types (see Section 2.4). Uninitialized integer variables have a default initial value of the largest negative integer, such as -2,147,483,647. See Variables later in this chapter (Section 2.3). There are two other integer *subtypes* defined in Package STANDARD, *positive* and *natural* (nonnegative); these are subsets of type integer.

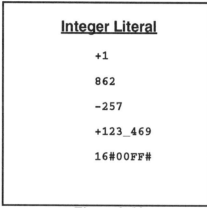

Figure 2-10

Unlike systems that represent an integer as a bit value, integers are not a vector of bits and cannot be indexed. Nor can you use logical operators on integers. Typically, conversion functions are used to convert an integer into a bit_vector when bit operations are needed. Vendors may also provide variations of integer, such as signed or unsigned bit_vectors, where arithmetic and logic functions are provided. You can specifiy the value of an integer in another radix, such as binary or hexadecimal. The last example indicates base **16** by using a **#**.

During design development, integers can be used as an abstraction of a signal bus or may represent an exact specification. Synthesis tools can usually build integer arithmetic blocks (adders, subtractors, comparators, etc.), which may default to 32-bit binary, unless a user specifies an explicit type (bit length). Some examples of length specification of a literal using a user-defined subtype are:

```
digit'(5)         word'(9)         I4'(6)
```

These examples depend upon the user defining what **word** and **digit** mean.

Time (Physical Literal)

The only predefined physical type is **time**, defined in Package STANDARD. *Physical literals* represent a unit of measurement. VHDL allows you to have a number and unit of measure, such as voltage, capacitance, and time. Any value specified by a physical literal is a multiple of the unit indicated. It is important to separate the number from the physical unit of measure with at least one space.

Examples:

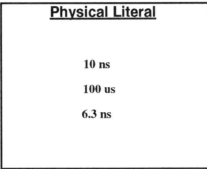

Physical Literal

10 ns

100 us

6.3 ns

Figure 2-11

Figure 2-12 defines the units of type **time**.

UNITS			
fs	=		femtosecond
ps	=	1000 fs;	picosecond
ns	=	1000 ps;	nanosecond
us	=	1000 ns;	microsecond
ms	=	1000 us;	millisecond
sec	=	1000 ms;	second
min	=	60 sec;	minute
hr	=	60 min;	hour

Figure 2-12

Physical literals are most useful in modeling and representing physical conditions during design testing. See *Figure 1-1* for an example using type **time**.

2.2 Names (Identifiers)

Schematics document the interconnection wires in electronic systems and associate symbolic names with wires. In VHDL, you also use symbolic names for signal wires. A name must begin with an alphabetic letter (a - z), followed by a letter, an underscore, or a digit. VHDL is not *case sensitive*. This means VHDL does not distinguish x from X. For example:

> **xyz** is recognized the same as **xYz**

The underscore, however, is significant, so:

> **x_3** is not the same as **x3**

Named objects include scalers, arrays, architecture names, process names, and entity names. Two different objects cannot have the same name (within the same scope). For example, two entities in a library cannot have the same name. Two architectures of a single entity each need unique names, and processes within an architecture need unique names. Array names can have an *index expression*. VHDL has over 100 reserved words that may not be used as identifiers. (See *Appendix A*)

Identifiers		**Invalid Names**
XYZ		in
S(3) Array Element		out
S(1 to 4) Array Slice		signal
X3		port
my_defs.unit_delay		library
		map
		x__3

Figure 2-13

Names can be relative; selected names indicate a package or a library:

> library_name.package_name.item_name
>
> library_name.item_name
>
> package_name.item_name

For example, in *Figure 1-1* the name is explicit:

> WORK.my_defs.unit_delay

2.3 Object Declarations

VHDL specifies electronic systems. In VHDL, a named signal may represent a wire in a physical design. During simulation, a signal value is stored in the memory of the host computer. Named objects are either constant (like ground) or varying in value.

Unlike programming languages, VHDL has two elements that can vary: the *variable*, which behaves just like a programming-language variable, and the *signal*, which is assigned a value at some specific simulated time. Because VHDL is a strongly typed language, variables and signals must be declared to be a specific type that never changes. The value may change, but the type never does.

The type must previously be declared. In the Package STANDARD the type is always visible, but in IEEE_1164, the type must be made "visible":

```
Library IEEE;
Use IEEE.STD_LOGIC_1164.all;
```

An object declaration statement declares a name and a type. There are four kinds of object declarations:

object_declaration   constant_declaration

                                                    signal_declaration

                                                    variable_declaration

                                                    file_declaration

Chapter 3 emphasizes the use of variables to facilitate understanding. Files are described in Chapter 4. Signals are discussed in Chapter 5.

Range Constraint

VHDL is a strongly typed language.   During simulation, signal and variable assignments are checked for type and *range*. Range constraints allow you to express design intent and scope of use. For example, you may want to refine an abstract representation of a counter to a counter that counts from 1 to 10. A range constraint declares the valid values for a particular type of signal or variable assignment. You can specify the range when you declare the variable or signal. The example below uses a range constraint in the port declaration to limit the values of **A** and **B**.

```
entity COMPARE_digit is

port (B,A: in integer range 0 to 9;              Range Constraint

    C:out boolean);

end COMPARE_digit;
```

Figure 2-14

A range constraint uses values, which should be compatible with the type it constrains, and be in a compatible direction with the original declaration of the type. Examples:

```
        integer range 1 to 10     NOT       integer range 10 to 1

        real range 1.0 to 10.0    NOT       real range 1 to 10
```

In some cases, you may want the entire range. For example:

```
        integer
```

The syntax is:

range_constraint

```
        range   index_constraint
```

index_constraint

```
        {low_val to high_val | high_val downto low_val}
```

Some declarations, such as variables and signals permit a range constraint as part of the declaration (e.g., [constraint]). Some array declarations permit an index constraint.

Constant Declaration

A constant is a name assigned to a fixed value. The name should be chosen to be meaningful to the reader. It is used in lieu of a literal, for convenience. If you need to make a change to the value of a constant (for example, in a subsequent refinement of a design), you need only change the constant declaration in one place. A constant makes a design more readable and makes updating code easy. A constant consists of a name, a type, and a value. It must be a valid literal for that type (see Section 2.1) and be in the correct range.

constant_declaration

Scalar: constant name: type :=expression;

Array: constant name: array_type [(index_constraint)] :=expression;

Examples:

```
   Constants

constant Vdd: REAL   :=   -4.5;

constant CYCLE: TIME := 100 ns;

constant PI: REAL := 3.14;

constant FIVE: INTEGER := 3 + 2;

constant FIVE:   BIT_VECTOR := "0101";              Length 4

constant FIVE:   std_logic_vector (8 to 11) := "0101";
```

Figure 2-15

Constants can be declared in a package, in a design entity, an architecture, or a subprogram. Frequently used or shared constants should be declared in a user-defined package (see Section 8.1 and *Figure 1-1*).

A constant specification in VHDL can also be used to specify a permanent electrical signal in connection to ground (or power) in a digital circuit.

Signal Declaration

Signals connect design entities together and communicate changes in values between processes. Signals can be abstractions of physical wires, busses, or used to document wires in an actual circuit. A signal with a type must be declared before the signal is used. The syntax is:

signal_declaration

Scaler: **signal** name(s): type[range_constraint] [:=expression];

Array: **signal** name(s): array_type[index_constraint] [:=expression];

Entity: **port** (names(s): direction type [range_constraint] [:=expression]);

Signals can be declared in several places (in an entity, in an architecture, or in a package). If you want to initialize a signal, indicate a literal in [*:=expression*] For example:

```
signal  S:  BIT:= '1';
```

Otherwise, the default initial value is the lowest value of that type.

```
  Signals

signal count: integer range 1 to 50;          Initial default is 1

signal GROUND: BIT:='0';

signal SYS_BUS std_logic_vector (7 downto 0);

port (B, A : in integer range 0 to 9);

signal bogus: bit_vector;                      Error no length
```

Figure 2-16

In VHDL simulation, signal assignments are scheduled in simulated time. For example:

```
signal xyz:BIT;

    xyz <= '1' after 5 ns;
```

This means **xyz** is set to **'1'** 5 ns after the time the simulator executes the statement. Signals cannot be declared in a process. They provide inter-process communication in an architecture or entity. Signals can be used in a process, but signal assignments within a process may cause unexpected results because the assignment of the value is delayed until a WAIT is executed (see Chapter 5).

Variable Declaration

A *variable* is a name assigned to a changing value within a process. A variable assignment occurs immediately in simulation, as opposed to a signal that is scheduled in simulated time. A variable can be used as a temporary simulation value, or to document a physical wire. A variable must be declared before it is used. The variable declaration must declare a type. The syntax is:

variable_declaration

Scaler: `variable` name(s): type [range_constraint] [:=expression];

Array: `variable` name(s): array_type [index_constraint] [:=expression];

A variable can be given a range constraint. It can also be given an initial value by including a literal value or [:=*expression*]. For example:

```
variable COUNT: INTEGER range 0 to 99 := 0;
```

The initial value of **COUNT** is **0** and can only be assigned integer values **0 to 99**. The initial value, by default, is the lowest (leftmost) value of range for that type.

```
     Variable Declarations


   variable INDEX:INTEGER range 1 to 50;

   variable CYCLE_TIME:TIME range 10 ns to 50 ns:=10 ns;

   variable MEMORY:BIT_VECTOR (0 to 7);   Index constraint

   variable x, y: INTEGER;                Defaults To -2,147,483,647
```

Figure 2-17

Local variables are *scalars* or *arrays* that can be declared in a process or a subprogram. Variables declared in a process are local data storage during simulation of a process or subprogram. For example:

```
variable x, y: INTEGER;

x := y + 1;
```

declares **x** and **y** to be integer variables that occupy simulator storage locations. The assignment statement changes the value of **x** to the current value of **y + 1** when simulated. As a hardware description, the output of the incrementer drives a value out to a wire named **x**.

In VHDL 92, global variables can be used to communicate between processes. (For more information, see Section 5.2 on Process Communication.)

2.4 Expressions

An expression is a formula that uses operators and defines how to compute or qualify a value. The operators must perform a calculation compatible with its operands. You must choose valid operands for an operator (see *Figure 2-19*). Generally, beyond the standard data types, vendors provide extended data types and overloaded operators (functions), as declared in a package (see Chapter 8). Vendors may provide operators with automatic type conversion, but the Package STANDARD does not.

Typing

Generally, operands must be of the same type. No automatic type conversion is done for you. For example, you *CANNOT* specify:

```
1 + 1.0
```

The result of an expression has a type that depends upon the types of operands and operators. For example, the result type of:

```
1 + 1
```
is INTEGER

and

```
1.0 + 1.0
```
is REAL

In an expression, an operand can be a name, a numeric, or a character literal (it can also be a function call, qualified expression, type conversion, etc.). Type conversion functions are built-in and provide a convenient way to change the type. The result type name is the name of the function.

```
+--------------------------------------------------------------+
|              Type Conversion Examples                        |
|                                                              |
|   integer (3.0)                       type integer           |
|                                                              |
|   real (3)                            type real              |
|                                                              |
|   integer * time                      type time              |
|                                                              |
|   nanos + picos                       type time              |
|                                                              |
|   nanos/picos                         type integer           |
|                                                              |
+--------------------------------------------------------------+
```

Figure 2-18

Below is a summary of operators. These operators create expressions that can calculate values. There are four kinds of operators: logical, relational, arithmetic, and concatenation.

		Operators		Operands
recedence	*logical_operators*	and	Logical And	same type
owest		or	Logical Or	same type
		nand	Complement of And	same type
		nor	Complement of Or	same type
		xor	Logical Exclusive Or	same type
	relational_operators	=	Equal	same type
		/=	Not Equal	same type
		<	Less Than	same type
		<=	Less Than or Equal	same type
		>	Greater Than	same type
		>=	Greater Than or Equal	same type
	concatenation_operator	&	Concatenation	
	arithmetic_operators	+	Addition	same type
		-	Subtraction	same type
	arithmetic_operators	+	Unary Plus	any numeric
		-	Unary Minus	any numeric
	arithmetic_operator	*	Multiplication	same type
		/	Division	same type
		mod	Modulus	integer
		rem	Remainder	integer
	arithmetic_operator	**	Exponentiation	integer exp.
		abs	Absolute Value	any numeric
ighest	*logical_operators*	not	Complement	same type

Figure 2-19

Logical operators work on predefined types bit, Boolean, std_logic, vectors of equal length, and not on integers. Do not mix them. The resulting expression has the same type as the type of the operands.

Relational operators compare two operands of the same type and produce a Boolean. The operands of the `<`, `>`, `<=`, and `>=` relational operators may be any scalar or one-dimensional array type whose element type is a discrete type (enumeration or integer). The result of an expression formed with a relational operator is of type Boolean.

Question: What does this mean?

```
X <= A <= B;
```
Answer:

Arithmetic operators work on integer, real, and std_logic_vector. If one or both of the operands of an arithmetic operator is not a literal, then the resulting expression has the type of that operand. However, the division of one physical type by the same physical type results in an expression of an integer type. Any physical type may be divided by an integer literal or floating point literal. Any physical type may be divided by the same physical type.

In some cases operators are specifications for a hardware block to be built using logic synthesis tools. Integer addition (+) produces a 32-bit adder unless a user specifies an explicit type (length). For example:

```
        variable a, b, x : integer range 0 to 255;

        x := a + b;
```

The example represents an 8-bit adder with wires **a**, **b**, and **x**, which can also be simulated using the current values of **a** and **b**.

Concatenation is defined for characters, strings, bits, and bit vectors and for all one-dimensional array operands. The concatenation operator **&** builds arrays by combining the operands. Each operand of **&** can be an array or an element of an array. Use **&** to append a single element to the beginning or end of an array, to combine two arrays, or to build an array out of elements. For example:

```
        "ABC"   &     "xyz"              results in "ABCxyz"

        "1001"  &     '0'                results in "10010"
```

Precedence

Operators in each box in *Figure 2-19* all have the same precedence level. The boxes in the table are in increasing order of precedence. For example:

```
A + B * C
```

means first multiply `B * C,` because multiply has a higher precedence level than add. You can use parentheses to override the default precedence. For example:

```
(A + B) * C
```

means do addition first.

You should use parentheses for operators of the same precedence level, or to override the order. For example:

```
(A and B) or C
```

Expressions

```
state3 and not reset

first_name & last_name

3 ns + 5 ps

(9/5) * c + 32

B >= 15

(HIGH + LOW)/2

(a and b) or c
```

Figure 2-20

Note: Parentheses can change the meaning of an array expression in subtle ways. For example:

`'1' & '0'` is equivalent to "10" , but `('1' & '0')` is not the same.

Generally the length of an array is lost when it is placed within parentheses.

IEEE Std_Logic_Vector Arithmetic

Recall that the Package STANDARD does not permit arithmetic on bit_vectors. Some vendors provide an arithmetic package that works with std_logic_vectors. Therefore, generally std_logic_vector is the most useful data type for synthesis and simulation. However, there is an ambiguity in interpreting the value of an std_logic_vector. For example:

```
"1011"
```

can be interpreted as either an unsigned value 11 or a signed value of -5, using 2's complement notation. Accordingly, some vendors provide two different arithmetic packages, which treat std_logic_vectors as either signed or unsigned. You need to declare which arithmetic package you want to access. For example:

```
use IEEE.std_logic_signed.all;        OR

use IEEE.std_logic_unsigned.all;
```

Figure 2-21

These packages define logical, relational, and arithmetic operators. For example:

```
"1011" > "0011"
```

The expression has two interpretations, depending upon your package selection:

```
11  > 3        Unsigned Numbers           (true)

-5 > 3         Signed Numbers             (false)
```

When using literal arithmetic expressions, the subtype name **signed** and **unsigned** can be used to indicate the unique type of operator desired. For example:

```
Signed'("1011") > Signed'("0011")
```

Since both operands are designated to be signed, the comparison operator (>) is **signed**, and produces a **false** condition.

New VHDL '92 Operators

sll	shift left logical	**ror**	rotate right
sla	shift left arithmetic	**srl**	shift right logical
rol	rotate left	**sra**	shift right arithmetic
xnor	exclusive nor		

- **Summary**

1. VHDL is a *strongly typed* language. The integer 1, the real number 1.0, and the bit '1' are not the same in VHDL. (Section 2.0)

2. One of the elements of VHDL is a *scalar*, which may be a named object or a literal made up of characters or digits, as long as it contains only one element. *Literals* are case sensitive. (Section 2.1)

3. Scalars may be of type *character, bit, boolean, real, integer,* or *physical quantity* defined in Package STANDARD. Vendor packages are included to extend the types available in VHDL. (Section 2.1)

4. In VHDL, you use symbolic names for a wire. A name must begin with an alphabetic character, followed by a letter, underscore, or digit. For names, VHDL is not *case sensitive.*(Section 2.2)

5. A named object may be a *constant* (unchanging value), a *variable* (changing value), or a *signal* (simulator time to schedule action). (Section 2.3)

6. *Expressions* are formulas that define how to compute or qualify a value. (Section 2.4)

User Package	**User Library**
Vendor Package	**Vendor Library**
Package TEXTIO	**Library STD**
Package STANDARD	
VHDL Language	**Library WORK**

Answers to Questions

Question: What does this mean?

```
X <= A <= B;
```

Answer: It means compare **A**, **B** to produce a Boolean result to assign to signal **x**.

-   **Exercise 1**

Type Conversion

Write a statement to convert **BIT** to **BOOLEAN** ('1' becomes **true**).

```
variable BITTY:  BIT;

variable BOOLY:  BOOLEAN;
```

Hint: Use the equality relation to produce a Boolean.

```
                    BOOLY  :=
```

Data types must match in an assignment statement.

- **Answer to Exercise 1**

Type Conversion

Write a statement to convert **BIT** to **BOOLEAN** (`'1'` becomes **true**).

```
variable BITTY:  BIT;

variable BOOLY:  BOOLEAN;
```

Hint: Use the equality relation to produce a Boolean.

```
BOOLY  :=   (BITTY = '1');
```

Data types must match in an assignment statement.

You need to do a bit-to-boolean conversion using a relation.

(**BITTY** = `'1'`) is relational and generates a **true** or **false**. (If BITTY = `'1'` then = true, and if BITTY = `'0'` then = false.)

- **Exercise 2**

Variable and Signal Exercise

What kind of VHDL object is each **bold** item in this example?

```
ENTITY and5  IS

    PORT (a, b, c, d, e : IN BIT;

        q :   OUT BIT);

END and5;

ARCHITECTURE behave OF and5  IS

BEGIN

    PROCESS (a, b, c, d, e)

        VARIABLE  state:  BIT;

        VARIABLE  delay:  TIME;

    BEGIN

        state := a AND b AND c AND d AND e;

        IF (state = '1')  THEN

        delay := 4.5 ns;

        ELSIF (state = '0')  THEN

        delay := 3 ns;

        ELSE

        delay := 4 ns;

        END IF;

        q <= state AFTER delay;

    END PROCESS;

END behave;
```

- **Answer to Exercise 2**

Variable and Signal Exercise

What kind of VHDL object is each **bold** item in this example?

```
ENTITY and5  IS
   PORT (a, b, c, d, e : IN BIT;
        q :  OUT BIT);                    output signal type bit
END and5;
ARCHITECTURE behave OF and5  IS           name of architecture
BEGIN
   PROCESS (a, b, c, d, e)
        VARIABLE state:  BIT;             variable type bit
        VARIABLE delay:  TIME;            variable type time
   BEGIN
        state := a AND b AND c AND d AND e; input signal
        IF (state = '1')  THEN
        delay := 4.5 ns;                  variable & physical type time
        ELSIF (state = '0')  THEN         literal type bit
        delay := 3 ns;
        ELSE
        delay := 4 ns;                    physical literal type time
        END IF;
        q <= state AFTER delay;           signal & variable type time
   END PROCESS;
END behave;
```

- **Exercise 3**

4-Input AND Gate Exercise

Assuming:

```
variable W, X, Y, Z, C: Bit;
```

Define a 4-input AND gate with inputs **w**, **x**, **y**, **z** and output **c**:

```
C :=
```

- **Answer to Exercise 3**

4-Input AND Gate Exercise

Assuming:

```
variable W, X, Y, Z, C: Bit;
```

Define a 4-input AND gate with inputs w, x, y, z and output c:

```
C := ( W and X and Y and Z);
```

- **Exercise 4**

Valid Assignment Statements Exercise

Which of the following statements are invalid?

```
variable A, B, C, D: BIT_VECTOR (3 downto 0);

variable E, F, G: BIT_VECTOR (1 downto 0);

variable H, I, J, K: BIT;
```

```
1) A := B xor C and D;

2) H := I and J or K;

3) A := B and E;

4) H := I or F;
```

- **Answer to Exercise 4**

Valid Assignment Statements Exercise

Which of the following statements are invalid?

```
variable A, B, C, D: BIT_VECTOR (3 downto 0);

variable E, F, G: BIT_VECTOR (1 downto 0);

variable H, I, J, K: BIT;
```

1) A := B xor C and D;	Invalid-Needs Parentheses
2) H := I and J or K;	Invalid
3) A := B and E;	Invalid-Mixes Types
4) H := I or F;	Invalid

- **Lab #2**

Part 1 - Monitoring simulation

1. Edit the source file below and run the analyzer `vhdlan a.ctof.vhd`.
    Run the simulator `vhdlsim ctof`.

2. Set a breakpoint on `f`.

3. Type `run 1` to start simulation and monitoring.

Part 2 - Using two algorithms

1. Copy the source file: `cp a.ctof.vhd b.ctof.vhd`.

2. Edit this new source file to add the algorithm ($g=2*c+30$).

3. Rerun the simulation as above but also monitor variable `g`.

4. Compare the results, make note here of accuracy:

C	f	g
0		
20		
40		

Source File `a.ctof.vhd`:

```
entity ctof is
end ctof;

architecture first of ctof is
begin
    a: process
    variable c, f, g: real;
    begin
    c := 0.0;
    while c < 40.0 loop
        f := 1.8 * c +32.0;
        c := c + 2.0;
    end loop;
    wait for 1 ns;
    end process;
end first;
```

3 | Sequential Statements

VHDL provides *concurrent statements* to document parallel operations or to abstractly model an ultimate circuit in a behavioral manner. These statements can be executed by a simulator at the same simulated time. The PROCESS statement is the primary concurrent statement in VHDL. Within a process, *sequential statements* specify the step-by-step behavior of the process. The dataflow style (shown in Chapter 6) is a shorthand form for the VHDL process.

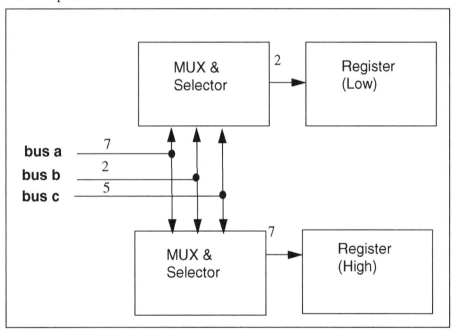

Figure 3-1

A number of different processes may run at the same simulated time. Consider the subsystem in the figure above, which captures the low value (2) and high value (7) from the three busses **a**, **b**, and **c**. The parallel abstract behavior of the system can be represented using two VHDL processes prior to deciding how the detailed function will be implemented in hardware.

In *Figure 3-1*, there are two registers that capture the `low` and `high` values on the three busses `a`, `b`, and `c`. *Figure 3-2* abstractly models the behavior using two processes `L` and `H` in an architecture:

```
Entity low_high is
   port (a, b, c: in integer);
end low_high;

architecture behavior of low_high is
begin

L: process
variable low: integer := 0;
begin
   wait on a, b, c;
   if a < b then low := a; else low := b; end if;
   if c < low then low := c; end if;
end process;

H: process
variable high: integer := 0;
begin
   wait on a, b, c;
   if a > b then high := a; else high := b; end if;
   if c > high then high := c; end if;
end process;

end behavior;
```

Figure 3-2

These processes, `L` and `H`, are two concurrent operations that identify the low and high values of 3 input integer signals `a`, `b`, and `c`, as shown in *Figure 3-1*. Each process begins with a **WAIT ON** `a, b, c`. Whenever `a` or `b` or `c` changes, both processes are run by the simulator. You do not know which process actually runs first, but as soon as one of the processes reaches the end, it loops to the top, reexecutes the `wait`, and stops; then the other process begins. The order of execution does not depend upon the order in which they are written (for example, process `H` may be simulated first). When the second process stops at its WAIT statement, both processes are again waiting for a change in the input. Although these processes actually are run sequentially, they are run at the same point in simulated time (for example, it could be at 3525 ns in simulated time). Also, they consume zero simulated time. Variable assignments and IF statements do not utilize simulated time. They are not scheduled events; instead, they are executed in a program-like manner by the simulator.

In this example, the architecture does not output any values; there are no outputs on the entity. Also, the values of local variables `high` and `low` are not available outside of the processes.

A process contains sequential statements that describe the behavior of an architecture. Sequential statements define algorithms for the execution within a process or a subprogram. These types of statements are the familiar notions of sequential flow, control, conditionals, and iterations.

Sequential statements execute in the order in which they appear in the process, as in programming languages. This chapter discusses the process, the sequential statements that may appear in a process, and the subprogram as follows:

- PROCESS Statement
- Variable Assignment Statement
- IF Statement
- CASE Statement
- LOOP Statement
- WAIT Statement
- NEXT Statement
- EXIT Statement
- Subprograms
- ASSERT Statement

3.1 PROCESS Statement

The PROCESS statement is a concurrent statement that defines the scope of each process. A PROCESS statement delineates the part of an architecture, where sequential statements are executed (components are not permitted in a process). This style of design is used for behavioral descriptions. The PROCESS statement provides "programming-language-like" capability using temporary *variables* that are declared inside the process. The syntax is:

```
process_statement
   [label :]
      process [(sensitivity_list)]
          [subprogram]
          [type]
          [constant]
          [variable]
          [other declarations]
      begin
          sequential_statements
      end process [label];
```

Figure 3-3

A process is an infinite loop that never exits; see Section 5.9 for more information about a process simulation cycle. The sensitivity list or a WAIT statement can be used to trigger a process execution. Declarations do not include signals. Variables declared are local to this process.

Sequential statements are logical, arithmetic, procedure calls, CASE statements, IF statements, loops, and variable assignments. These statements can only be used inside a process body or a subprogram. The process_label is useful in some simulators for debugging (for example, setting a breakpoint).

3.2 Variable Assignment Statement

A variable assignment statement replaces the current value of a variable with a new value specified by an expression. The named variable and the result of the expression must be of the same type. The syntax is:

variable_assignment_statement

target := expression;

The left side (target) of the variable assignment statement is a variable previously declared. The right side of the variable assignment statement is an expression using variables, signals, and literals. This statement executes in zero simulation time and acts similar to assignments in most programming languages. Assignments can be made to a scaler or to an array.

For example:

```
ix := 'a';                Character

a := 1.0;                 Real assignment
```

The following are more examples:

```
A := 1;

B := 2;

C := A + B;               The value of C is 3
```

Recall a signal assignment uses the operator <=.

Note: Variables declared within a process cannot pass values outside of the process; that is, they are *local* to a process or subprogram (see Section 2.3).

VHDL does not support redefinition (overloading) of the := assignment operator. (See Operator Overloading in Chapter 9.)

3.3 Sequential Signal Assignment Statement

A signal assignment evaluates the right side expression and causes the simulator to
schedule an update of the target. The scheduling can only occur after the simulator
executes a WAIT statement (see Chapter 5). The syntax is:

signal_assignment_statement

target <= [expression] [**after** delay];

3.4 IF Statement

IF statements represent hardware decoders in both abstract and detailed hardware
models. The IF statement selects for execution one or more of the enclosed
sequences of statements, depending upon the value of one or more corresponding
conditions. The syntax is:
if_statement

if condition **then** sequential_statements

{**elsif** condition **then** sequential_statements}

[**else** sequential_statements]

end if;

There are three examples of the IF statement: the simple IF..THEN, the
IF..THEN...ELSE, and the IF..THEN...ELSIF. For example:

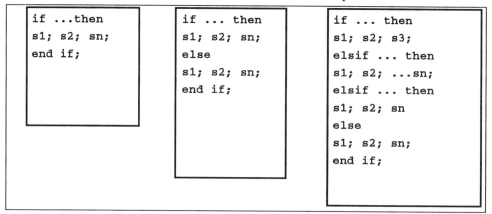

Figure 3-4

The first example is a simple IF...THEN, which tests some condition and executes
a series of statements labeled **s1**, **s2**, up to **sn**. A more complex statement is the
IF..THEN..ELSE statement where a condition is evaluated, and either one set of
statements is executed or the other set of statements is executed, depending upon

the truth of the condition being tested. A   more complex case is the
IF..THEN..ELSIF. This is a nested IF statement where the first condition is
evaluated. If the first condition is false, the ELSIF is evaluated; and if it is true, the
statements following are executed. Note that you can have a number of nested
ELSIF statements and you can have (optionally) an ELSE statement which goes
with the original IF statement. If none of the statements are executed, then the final
ELSE statement is executed.

Following is a simple example from an Official Airline Guide, which has codes for
the days of the week. 1 is Monday, 6 is Saturday, and 7 is Sunday.

```
IF (day = 7)    THEN weekend := TRUE;
ELSIF (day = 6) THEN weekend := TRUE;
ELSE  weekend := FALSE;
END IF;
```

In the following example, the last assignment to T has highest priority.

```
IF (X) then T:=A; end if;
IF (Y) then T:=B; end if;
IF (Z) then T:=C; end if;
```

is equivalent to:

```
IF (Z) then T:=C;
   elsif (Y) then T:=B;
      elsif (X) then T:=A;
end IF;
```

3.5 CASE Statement

CASE statements are useful to describe decodings of busses and other codes. The CASE statement selects, for execution, one of a number of alternative sequences of statements. The chosen alternative is defined by the value of an expression.

Use the CASE statement in VHDL when you have a complex decoding situation. It is a more readable statement than a collection of nested IF statements. It allows you to easily identify the value and associated actions. The syntax is:

case_statement

case expression **is**

    when choices-1 => sequence_of_statements

        .

        .

    when choices-n => sequence_of_statements

end case;

The example shows converting **bitindex**, type bit, to an integer variable **ix**.

```
CASE bitindex IS

    WHEN '0' => ix := 0;              => Means "Then"

    WHEN '1' => ix := 1;

END CASE;
```

Figure 3-5

You can read the example as: when **bitindex** is a '0' bit, then set **ix** to an integer 0; when **bitindex** is a '1', then set **ix** to an integer 1. The conversion might be needed because **bitindex** is of type bit and an integer might be needed in the application.

The following are options of the CASE statement:

```
case expression is

   when value => s1;s2;..sn;                    => means "Then"

   when val1 | val2| valn => s1;s2;..sn;        | means "or"
              .
   when val1 to val2 => s1;s2;..sn;             range
              .
              .
   when others => s1;s2;..sn;                   others (default)

end case;
```

Figure 3-6

The CASE statement contains multiple WHEN clauses. WHEN clauses allow you to decode particular values and enable actions **s1**, **s2** following the right arrow (**=>**). Usually, you have a set of values to test and several actions to do (first WHEN clause in *Figure 3-6*). You can also have single *discreet* values that are tested (indicated by the "or" indicator | shown in the second WHEN clause). You can have a range of values **when val1 to val2**. The OTHERS clause takes all other cases if none preceeding are true, to cause some action to occur. **s1; s2; .....sn;** are sequential statements that can include the NULL statement, **null**. Unlike some programming languages, you do not need a BEGIN and END around a group of statements (**s1; s2; ...sn**). The END CASE terminates the statement.

Rule: The CASE statement selects a number of actions. It must enumerate all possible values of expression or have an OTHERS clause. The OTHERS clause must be the last choice of all of the choices.

This example documents a BCD to 7-segment decoder circuit's behavior:

```
case BCD is
   when "0000" => LED := "1111110";
   when "0001" => LED := "1100000";
   when "0010" => LED := "1011011";
   when "0011" => LED := "1110011";
   when "0100" => LED := "1100101";
   when "0101" => LED := "0110111";
   when "0110" => LED := "0111111";
   when "0111" => LED := "1100010";
   when "1000" => LED := "1111111";
   when "1001" => LED := "1110111";
   when others => LED := "-------";  --don't care
end case;
```

Figure 3-7

3.6 LOOP Statement

LOOP statements provide a convenient way to describe bit-sliced logic or iterative circuit behavior. A LOOP statement includes a sequence of statements to be executed repeatedly, zero or more times. The syntax is:

loop_statement

> [label:] [**while** condition | **for** loop_specification] **loop**
>   sequential_statements
> **end loop** [label];

LOOP statements are common to many programming languages. The LOOP statement is a sequential statement in a process. It can optionally have a label, which is useful in nested loops. There are two different styles of the LOOP statement: the FOR LOOP and the WHILE LOOP. Examples:

```
L: FOR i IN 1 TO 10 LOOP
    s1; s2; ..sn;                    Sequential Statements
END LOOP;
          and

i := 1;
M: WHILE (i <11)  LOOP
       s1,s2,..sn;                   Sequential Statements
       i := i + 1;
END LOOP;
```

Figure 3-8

The first example is a FOR LOOP. It iterates 10 values for i, which is, by default, integer and takes on the values 1 to 10. The second example is a WHILE LOOP. If the condition **WHILE(i <11)** evaluates to **true**, it continues to loop. **end loop [label]** is required. Loops can be nested (see *Figure 3-10*). The label in a LOOP statement is optional. The LOOP identifier, i, in the example does not need declaration and is considered local to the loop. You cannot change its value from within the loop, nor use it outside of the loop. Although a PROCESS statement is effectively an infinite loop, for style reasons you may want to write the process as an infinite loop as follows:

```
process
begin
    initialization_statements
    loop
        sequential_statements;
    end loop;
end process;
```

Notice that the loop is not qualified or limited by a FOR or a WHILE. The loop must contain a WAIT statement (see Section 3.8).

3.7 NEXT Statement

The NEXT statement skips execution to the next iteration of an enclosing LOOP statement (called *label* in the syntax). The completion is conditional if the statement includes a condition. NEXT is convenient to use when you want to skip an iteration of a LOOP. The syntax is:

next_statement

> next [label] [when condition];

NEXT stops execution of the current iteration in the LOOP statement and skips to successive iterations. For example:

```
FOR i IN 0 TO max_limit LOOP
   IF (a(i) = 0) THEN NEXT;
   END IF;
q(i) := a(i);
END LOOP;
```

Figure 3-9

The LOOP statement has a range bounded by the END LOOP. Execution of the NEXT statement causes iteration to skip to the next loop index value (e.g., For i). The arrow in *Figure 3-9* indicates the scope of where execution jumps to.

When loops are nested, you may want to give each loop a label. The NEXT statement refers to a particular loop label. For example:

```
L1: WHILE i <10 LOOP
L2: WHILE j <20 LOOP

   .
   .
   .

      NEXT L2 when i = j;
   .
   .

   .
END LOOP L2;
END LOOP L1;
```

Figure 3-10

The NEXT can be conditional, as shown in *Figure 3-10*.

3.8 EXIT Statement

The EXIT statement completes the execution of an enclosing LOOP statement (called *label* in the syntax). The completion is conditional if the statement includes a condition.

exit_statement

        exit [label] [when condition]

EXIT stops execution of the iteration of the LOOP statement. For example:

```
FOR i IN 0 TO max LOOP
   IF (a(i) = 0)) THEN EXIT; END IF;
       q(i) := a(i);
END LOOP;
```

Figure 3-11

The FOR statement has a range for LOOP. IF(a(i) = 0), then EXIT causes execution to exit the loop entirely. The arrow indicates the scope of where execution jumps to.

The loop label in the EXIT statement identifies the particular loop to be exited. Nested loops should be given unique labels.

3.9 WAIT Statement

The WAIT statement provides for modeling signal-dependent activation. Conceptually, nothing happens in a physical flip-flop until a clock is activated. That is, the flip-flop is waiting for a clock until it uses the input signals. Use the WAIT statement to model a logic block that is activated by one or more signals. The WAIT statement causes a simulator to suspend execution of a process statement or a procedure, until some conditions are met. (It also forces simulator signal propagation - see Chapter 5.) The syntax is:

wait_statement

    wait   [on signal_names]
            [until conditional_expression]
            [for time_expression]

A WAIT statement can appear any place in a process and can appear more than once within a process, for different purposes. There are four types of WAIT statements: WAIT FOR, WAIT UNTIL, WAIT ON and WAIT.

Examples: **WAIT ON a,b;**

This example has a signal list that suspends execution until a change occurs on either signal **a** or **b**. When either signal changes, the **WAIT** is satisfied and process execution begins. However, a signal must have a *change event* such as from a Boolean **true** to a **false**. An assignment of a signal to the same value is not an event (change). In lieu of an explicit **WAIT ON** statement, a process sensitivity list can be used. (See Section 5.11.) For example:

Process (A, B);

The example below suspends execution of the process until condition **x > 10** becomes satisfied. It requires an event on **x** to evaluate the expression.

WAIT UNTIL x > 10;

The example below suspends execution of a process for 10 nanoseconds.

WAIT FOR 10 NS;

A process can be suspended indefinitely:

WAIT;   Waits Forever

3.10 ASSERT Statement

During simulation, it is convenient to output a text string message as a warning or error message. The ASSERT statement allows for testing a condition and issuing a message. The ASSERT statement checks to determine if a specified condition is true, and displays a message if the condition is false. The syntax is:

assertion_statement

assert condition [**report** string_expression] [**severity** expression];

ASSERT writes out text messages during simulation. There are four levels of severity: FAILURE, ERROR, WARNING, NOTE. The ASSERT statement is useful for timing checks, out-of-range condition, etc. For example:

```
ASSERT (x > 3)                    Prints if condition is false
REPORT "setup violation"
SEVERITY WARNING;
```

Figure 3-12

The severity level is used is the simulator to either terminate a simulation run or to just give a warning message and continue. Vendors provide various ways to handle the severity levels. To unconditionally print out a message, use the condition **false**. For example:

assert(false) report "starting simulation";

In VHDL 92 you can use the report as a statement without the assert (condition).

3.11 Subprograms

In behavioral design descriptions, subprograms provide a convenient way of documenting frequently used operations. A subprogram has sequential statements contained inside, and is *called* from a process. There are two different types of subprograms: a *procedure* (returns multiple values) and a *function* (returns a single value). Typical uses are *conversion functions* or *resolution functions*. Subprograms are either built-in or user-defined. See Section 9.2 for more information on resolution functions.

Note: Subprograms can declare local variables, but no value is remembered after the subprogram is exited. This behavior is similar to subprograms in Pascal.
During simulation debugging, it may not be possible to display a local variable within a subprogram if it is not active.

Functions

A user-defined function needs to be declared if it is called. When you call a function, values are passed in through parameters just prior to its execution. Parameters coming into the function are of input mode, and are passed as simple values. The function executes and returns only one value. This means that a function actually executes and evaluates like an expression. Nothing can be passed back through the parameters; you can only return a single value. Function may not contain a WAIT statement.
Below is an example of a function declaration and the call of a function `C_to_F`, which converts Celcius to Farenheit .

```
PROCESS
Function C_to_F (c: real) return real is
Variable F:real;                              Declaration
BEGIN
   F := c * 9.0/5.0 + 32.0;
   Return (F);
end C_to_F;
_____
Variable new_temp: real;
BEGIN
   new_temp := C_to_F (5.0) + 20.0;           Call
   .....
end process;
```

Figure 3-13

A user-defined function must be declared before you call it. In the declaration, `function` is a reserved word. You are declaring a function `C_to_F`. It brings in one real argument named `c`, the Celcius temperature, which is a *dummy variable* local to this function. The **return** value is a single value of type **real**. There is a declaration of a **variable F** used only in the function for temporary storage. Local variables are initialized each time the function is called; the function has no memory from call to call. The body of the function follows the **BEGIN** statement. It specifies that **c** (being passed in) is multiplied by 9 and divided by 5, then added to 32. Assigned as the value of **F**, the **return** value **F** passes back the Farenheit temperature.

In the example process, you call the function **c_to_F** (5.0), where the 5.0 is the actual parameter for **c**. After the function executes, it returns the value as if it were an expression. The left side of the plus sign has the value that is returned (41.0 in this case) when the function executes; the plus (**+**) adds 20 to calculate the new temperature, **new_temp** (61 in this case).

The next example shows where to declare a function and where to call the function. The function declaration goes in the process declaration section, between the word **process** and the **begin**. Here you declare functions used in this process. Calls to the function are sequential statements inside a process after the **begin**.

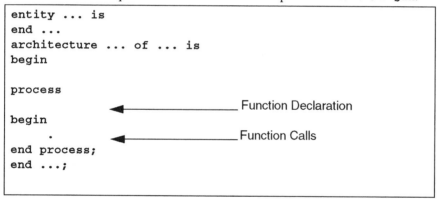

```
entity ... is
end ...
architecture ... of ... is
begin


process
                                        Function Declaration
begin
      .                                 Function Calls
end process;
end ...;
```

Figure 3-14

Functions can also be declared in an entity, and architecture, or a package. Many vendors provide utility functions for sin, cos, sqrt, etc. in a VHDL package. These are source code design units that you compile and use from your VHDL library (see Section 8.7).

A conversion function converts one type of a named variable to another. Conversion functions can be built-in or user-supplied. A designer uses a conversion function to use an object or entity from another design that is of a different data type. For example:

```
FUNCTION vect_to_int (s : bit_vector (1 to 8)) RETURN INTEGER IS
   VARIABLE result : INTEGER := 0;
BEGIN
   FOR i IN 1 TO 8 LOOP
   result := result * 2;
      IF s(i) = '1' THEN
      result := result + _____;              Fill in the Blank
      END IF;
   END LOOP;
   return result;
END vect_to_int;
```

Figure 3-15

This function returns an integer when you pass in a `bit_vector`. The function tests bits in an array `s`. If a bit is a `'1'`, it adds to the `result`. In this way, the function converts an 8-bit data `bit_vector` to unsigned integer.

Question: What value should be put in `result = result + _____;`?

Answer:

Following are examples of *built-in* conversion functions:

        `Integer (xyz)`   Argument is Real

        `Real (xyz)`   Argument is Integer

Both examples do type conversion. The first returns an integer for an argument, type `real`. The second converts an integer argument into type `real`.
A resolution function is not explicitly called (see Section 9.2).

Procedures

A procedure is also a type of subprogram. With a procedure you can return more than one value, using parameters. The parameters are of mode **IN**, **OUT**, and **INOUT**. If not specified, the default is **IN**. **IN** brings a value in, **OUT** sends a value back through an argument list, and **INOUT** brings a value in and sends it back. Parameters can be signals or variables. If not declared as signals, they default to variables. Procedures use the return statements without a return value, as functions do. The return value is passed via an assignment statement, you must choose a compatible variable assignment statement (**:=**) or signal assignment statement (**<=**) in a procedure. Procedures can contain WAIT statements, and signal parameters can pass signals to be waited on. Local variables can be declared in a procedure. Local variables are initialized each time the procedure is called; the procedure has no memory from call to call.

A procedure call is a statement (not an expression), which means you use a procedure call on a line by itself. The procedure must be declared in a package in a process header or in an architecture declaration prior to its call.

The syntax is:

procedure_declaration

```
procedure name (parameters) is
    [variables]
    [constants]
    [types]
    [declarations]
begin
    sequential_statements
end ;
```

parameters

```
{[variable] names [in | out | inout] type [:= expression]; |
    signal names [in | out | inout] type;}
```

Parameters can be assigned a default value that is used when no actual parameter is specified in a procedure call.

The example in *Figure 3-16* is a procedure declaration that converts a vector of bits into an integer. The procedure body shows assignment to parameters q and zero_flag to send back values, converting the bits in z to an integer q and indicating whether the result was 0. There are 3 parameters: z coming IN, zero_flag coming OUT, q coming INOUT.

```
PROCEDURE vector_to_int (z : IN bit_vector (1 to 8);
                         zero_flag : OUT BOOLEAN;
                         q : INOUT INTEGER) IS
BEGIN
   q := 0;
   zero_flag := TRUE; -- for zero
   FOR i IN 1 to 8 LOOP
      q := q * 2;
      IF (z(i) = '1') THEN q := q + 1;
   zero_flag := FALSE; -- for nonzero
   END IF;
   END LOOP;
   return;
END vector_to_int;
```

Figure 3-16

Question: Why is q INOUT?
Answer:

Question: When you call this procedure:

                vector_to_int (s, t, u);

show the declaration of t and u.

Answer:

The next example shows a procedure for an 8-bit parity generator.

```
procedure PARITY (A: in BIT_VECTOR (0 to 7);
                  RESULT1, RESULT2: out BIT) IS
variable TEMP: BIT;
begin
  TEMP := '0';
  for I in 0 to 7 loop
  TEMP := TEMP xor A(I);
  end loop;
  RESULT1 := TEMP;
  RESULT2 := NOT TEMP;
end;
```

Figure 3-17

It brings in an 8-bit vector **a** and sends back two values, the even and the odd parity. When you call this procedure you must use parameters that are variables (not signals) because the procedure does variable assignments to parameters **RESULT1** and **RESULT2**.

> `variable: x,y:bit;`   *NOT*   `signal x, y: bit;`

Procedures calls are positional or named association, `parameter => actual`:

> `PARITY (my_vector, x, y);`   *OR*

> `parity (A=> my_vector, result2 => y, result1 => x);`

The example in *Figure 3-17* shows two procedure calls to a previously declared procedure **PARITY**, which is a 8-bit parity checker. It is called twice to check the 16 bits of **Y**: first the upper 8 bits of **Y**, returning **TOP**; then the lower 8 bits of **Y**, returning the value back through the variable **BOTTOM**. The example shows the procedure calls, separated by semicolons, inside the main body of the process; the procedure declaration is not shown.

```
architecture BEHAVIOR of RECEIVER is
process
   variable TOP, BOTTOM, ODD, dummy: BIT;
   variable Y : bit_vector (15 downto 0);
begin
   .
   .
   .
   PARITY (Y(15 downto 8), TOP, dummy);   ◀──────Procedure calls
   PARITY (Y(7 downto 0), BOTTOM, dummy);  ◤
   ODD := TOP xor BOTTOM;
end process;
end BEHAVIOR
```

Figure 3-18

Question: Is there any significance of two calls to an 8-bit parity generator versus a 16-bit loop in parity generator?

Answer:

Question: Can you rewrite *Figure 3-18* without TOP, BOTTOM, and ODD?

Answer:

- **Summary**

1. A *process* defines regions in architectures where sequential statements are executed (components are not permitted). (Section 3.1)

2. Process statements provide concurrent processing capability using *local* variables and *global signals*. (Section 3.1)

3. VHDL contains *sequential statements*, including IF THEN ELSE, CASE, LOOP, etc. (Sections 3.2, 3.3, 3.4, etc.)

4. WAIT statements dynamically control process suspension/execution. In simulation, all processes are started and executed up to a WAIT. (Section 3.8)

5. A process can call *functions* (that return a single value) and *procedures* (that return more than one value). (Section 3.9)

Process

Declarations
Internal variables that hold temporary values in the sequence of computations, as well as types, constants, components, and subprograms used locally.

Sequential Statements

Signal assignments	**LOOP statements**
Compute values and assign them to signals.	Execute statements repeatedly.

Procedure Calls	**NEXT statements**
Invoke predefined algorithms.	Skip remainder of LOOP.

Variable statements	**EXIT statements**
Store partial results in variables.	Terminate the execution of a LOOP.

IF statements	**WAIT statements**
Conditionally execute groups of sequential statements.	Wait for a clock signal.

CASE statements	**NULL statements**
Select a group of sequential statements to execute.	Are place-holders that perform no action.

- Answers to Questions

Question: What value should be put in `result = result + _____`?

Answer: `result = result + 1`

Question: Why is `q` INOUT?

Answer: Because it is used on the left- and right-hand side of the assignment statement.

Question: When you call this procedure:

```
vector_to_int (s, t, u);
```

show the declaration of `t` and `u`.:

Answer:

```
variable t: Boolean;
```

```
variable u: integer;
```

Note that you cannot use signals.

Question: Is there any significance of two calls versus a 16-bit loop in parity generator?

Answer: Function is the same. In case of logic synthesis, resultant hardware implemented might be different speed/area.

Question: Can you rewrite *Figure 3-17* without TOP, BOTTOM, and ODD?

Answer: Yes, using a complex expression in the RETURN statement.

- **Exercise 1**

Write a loop to calculate the conversion of Celsius to Fahrenheit for 0 to 40° C.

```
Architecture
   begin
   process
   variable c, f: real;
   begin
```

- **Answer to Exercise 1**

Write a loop to calculate the conversion of Celsius to Fahrenheit for 0 to 40° C.

```
Architecture
   begin
   process
   variable c, F: real;
   begin


       C := 0.0;
       WHILE c < 41.0   LOOP
       F := c * 9.0/5.0;
       F := F + 32.0;
       C := C + 1.0;
       end loop;
       wait for 1 ns;
       end process;
```

Create a WHILE LOOP for the variable, to loop as long as c = 0 to 40 C. In the loop, multiply c by 9 and divide by 5, then add 32 degrees Fahrenheit. Reexecute the loop while the value of c is less than 41.

-   **Exercise 2**

The exercise below involves the function **shiftr**. It operates on an integer in the range of 0 to 255. The formal argument is called **a**. Calculate the average of **x, y** using the **shiftr** function.

```
FUNCTION shiftr (a : integer range 0 to 255) RETURN integer IS
   BEGIN
      RETURN (a/2);
   END shiftr;




      variable x, y, average: integer range 0 to 255.
```

Answer to Exercise 2

The exercise below involves the function **shiftr**. It operates on an integer in the range of 0 to 255. The formal argument is called **a**. Calculate the average of **x, y** using the **shiftr** function.

```
FUNCTION shiftr(a : integer range 0 to 255) RETURN integer IS
   BEGIN
      RETURN (a/2);
   END shiftr;
```

```
   variable x, y, average : integer range 0 to 255;
   average :=    shiftr(x) + shiftr(y);
```

Note that **shiftr (x + y)** may exceed the range 0 to 255.

- **Exercise 3**

Rewrite the function below without a local variable.

```
Function C_to_F (c: real) return real IS
Variable F:real;
BEGIN
F := c * 9.0/5.0;
F := F + 32.0;
Return (F);
end function;
```

- **Answer to Exercise 3**

Rewrite the function below without a local variable.

```
Function C_to_F (c: real) return real is
BEGIN
Return ((c * 9.0/5.0) + 32.0);
end function;
```

- **Exercise 4**

Drink Machine Exercise

In the example, consider coin inputs of 5, 10, 25. What states are entered and what
actions occur starting with `current_state = idle`?

```
case CURRENT_STATE is
  when IDLE =>                        when TEN =>
  if (FIVE_IN) then                   if (FIVE_IN) then
     NEXT_STATE <= FIVE;                 NEXT_STATE <= FIFTEEN;
     elsif (TWENTY_FIVE_IN) then         elsif (TEN_IN) then
     NEXT_STATE <= TWENTY_FIVE;          NEXT_STATE <= TWENTY;
     end if;                             elsif (TWENTY_FIVE_IN) then
                                         NEXT_STATE <= IDLE;
  when FIVE =>                           DISPENSE <= TRUE;
  if (FIVE_IN) then                      end if;
     NEXT_STATE <= TEN;
     elsif (TEN_IN) then              when FIFTEEN =>
     NEXT_STATE <= FIFTEEN;          if (FIVE_IN) then
     elsif (TWENTY_FIVE_IN) then         NEXT_STATE <= TWENTY;
     NEXT_STATE <= THIRTY;               elsif (TEN_IN) then
     end if;                             NEXT_STATE <= TWENTY_FIVE;
                                         elsif (TWENTY_FIVE_IN) then
                                         NEXT_STATE <= IDLE;
                                         DISPENSE <= TRUE;
                                         FIVE_OUT <= TRUE;
                                         end if;
                                         .
                                         .
```

     (5)   (10)   (25)

-   **Answer to Exercise 4**

Drink Machine Exercise

In the example, consider coin inputs of 5, 10, 25. What states are entered and what actions occur?

```
case CURRENT_STATE is
   when IDLE =>                           when TEN =>
   if (FIVE_IN) then                      if (FIVE_IN) then
      NEXT_STATE <= FIVE;                     NEXT_STATE <= FIFTEEN;
      elsif (TWENTY_FIVE_IN) then            elsif (TEN_IN) then
      NEXT_STATE <= TWENTY_FIVE;             NEXT_STATE <= TWENTY;
      end if;                                elsif (TWENTY_FIVE_IN) then
                                             NEXT_STATE <= IDLE;
   when FIVE =>                             DISPENSE <= TRUE;
   if (FIVE_IN) then                        end if;
      NEXT_STATE <= TEN;
      elsif (TEN_IN) then                when FIFTEEN =>
      NEXT_STATE <= FIFTEEN;             if (FIVE_IN) then
      elsif (TWENTY_FIVE_IN) then           NEXT_STATE <= TWENTY;
      NEXT_STATE <= THIRTY;                 elsif (TEN_IN) then
      end if;                               NEXT_STATE <= TWENTY_FIVE;
                                            elsif (TWENTY_FIVE_IN) then
                                            NEXT_STATE <= IDLE;
                                            DISPENSE <= TRUE;
                                            FIVE_OUT <= TRUE;
                                            end if;
                                               .
                                               .
```

                    states   actions

Answer: FIVE, FIFTEEN, IDLE, DISPENSE, FIVE_OUT

        (5)   (10)   (25)

- **Lab #3**

Using a common function

1. Utilize a function in the declarative part of the architecture that does the conversion algorithm of Celsius to Fahrenheit. Express the algorithm y = mx + b; with input parameters x, m, b. Call the function from the processes (inside a loop) that uses two calls to one function.

2. Analyze and simulate your design.

Using integer mode

3. Edit this design to use integer mode instead of real arithmetic, and use the g algorithm (since 1.8 is not integer) in one process a. Assume that all Celsius data is in the range 0 to 40.

4. Test your design with the analyzer and simulator.

Using two processes

1. Fix this file to contain one architecture with two separate processes. Each has a unique algorithm. You also need to change the process name in the second process from a to b. Calculate f in process a, and g in process b.

2. Analyze and simulate this design. Set the break points when f, g change. What is the order of process execution? How much simulated time is used by the two processes?

4

Advanced Types

This chapter introduces data types that are more advanced than the predefined types previously discussed in Chapter 2 (types *integer, real,* and *physical,* and predefined enumerated types *boolean* and *bit*). The advanced data types include *enumerated types* that allow for identifying specified values for a *type* and for *subtypes,* which are variations of existing types. There are composite types (as opposed to *scalar* types), that include *arrays* (vectors, as they are often known) and *records,* which are more complex. And there are predefined data types, *text* and *lines,* that facilitate text input and output operations. These types and functions are provided in Package TEXTIO, which is contained in library STD. This chapter describes advanced data types as follows:

- •Extended Types
  - Enumerated Types
  - Subtypes
- • Composite Types
  - Arrays
  - Records
- •Other Predefined Types
  - Files
  - Lines

The extensible type facility of VHDL is helpful in modeling behavior at a more abstract level and can also be used to specify very detailed behavior. Symbolic coding of values hides lower-level details, makes your code more readable, and allows you to defer the assignment of numeric values. It also allows you to define special types and avoid mismatching connecting components.

4.1 Extended Types

First-generation languages restricted the data types to a few predefined types, such as integers and real numbers. The VHDL Language does not include many built-in types for signals and variables, but allows vendors and users to add new data types. The Package STANDARD, included in every implementation, extends the language to allow interesting data types for the description of hardware. These types include:

- Boolean

- Bit

- Bit_vector

- Character

- String

- Text

For example, the Package STANDARD includes the following type declaration:

```
type boolean is (false, true);
```

The above enumerates the two possible values for type boolean. By default, any signal or variable defined of this type is checked for valid assigned values during simulation. The initial value of a signal or variable of this type is the lowest (leftmost) value **false**. In a similar way, another type declaration in the Package STANDARD is:

```
type bit is ('0', '1');
```

Additionally, the Package IEEE 1164 declares the type **std_logic** as:

```
type std_logic is ('U', 'X', '0', '1', 'Z', 'W', 'L', 'H', '-');
```

Since the **'U'** is declared as the leftmost element of this type, it is the default initial value for signals of this type and is interpreted by simulators as uninitialized.

To extend the available data types, VHDL provides a type-declaration capability and a package facility to make it convenient for vendors to deliver and users to use these new data types. VHDL also provides *overloaded* operators so that the use of these data types is natural and easy (see Section 9.1).

VHDL Package STANDARD does *not* provide for real arrays or for integer arrays (only bit vectors). Therefore, the user or vendor must declare new data types to implement vectors or matrix representations. You need to declare a type before you declare a signal or a variable of that type. The following section illustrates the capabilities in VHDL for enumeration and subtypes.

Enumerated Types

The enumerated type declaration lists a set of names or values defining a new type. The syntax is:

enumerated_type_declaration

     type identifier **is** (item {, item});

     item
           { identifier | character_literal}

This feature allows you to declare a new type using character literals or identifiers. For example:

using identifiers   **TYPE tools IS (hammer, saw, drill, wrench);**

The example identifies 4 different values in a particular order that define type **tools**. In subsequent declarations of a variable or signal designated type **tools**, assigned values could only be **hammer, saw, drill, wrench**.

In your application, you may find it convenient to represent codes symbolically by defining your own data type, such as **tools** above, using identifiers.

Another example using character literals is:

using literals   **TYPE fiveval IS ('?', '0','1','Z','X');**

This example is particularly interesting for logic simulation because it has delineated five values **'0'**, **'1'**, **'X'**, **'Z'** and **'?'**. The value **'?'** is declared as the leftmost value to be the default simulation value for uninitialized signals. These values are meaningful in electrical engineering and are an extension of **'1'** and **'0'**, the two standard values for a bit. If you had declared type **fiveval** in a design, you could declare ports, signals, and variables of this type.

Package IEEE 1164 declares std_logic as probably the most popular simulation strengths. Previously, vendors provided data types such as multivalued logic (**MVL7** and **MVL9**).

Below is an example using an enumerated type. It declares type `Instruction`.

```
ARCHITECTURE behave OF cpu IS
TYPE Instruction IS (add, Lda, Ldb);
BEGIN PROCESS
VARIABLE a, b, data: INTEGER;
VARIABLE Instruct: Instruction;
BEGIN
.........
CASE Instruct IS
   WHEN Lda => a:= data;          Load a accumulator
   WHEN Ldb => b:= data;          Load b accumulator
   WHEN add=> a:= a + b;          Add two accumulators
END CASE;
wait on data;
END PROCESS;
END behave;
```

Figure 4-1

In *Figure 4-1*, the CASE Instruct statement is discriminating on the values of the variable Instruct declared of type Instruction. The only values that Instruct can take on are the enumerated values of the enumerated type Instruction. In the CASE statement, *Figure 4-1* tests the variable Instruct for the values Lda, Ldb, and add. Each of the corresponding actions depend upon the value of the Instruct variable, which can be only one of those enumerated values.

Enumerated types also provide for abstraction and information hiding. They allow more design errors to be detected during the compilation process and can reduce debugging time. Additionally, enumerated types provide a more abstract design style that computer scientists refer to as *object oriented*. Enumerated types allow you to use symbolic values instead of numeric codes. As the designer, you do not have to specify real numeric codes. If logic is synthesized, a numeric encoding is generated automatically by software. Some extensions to VHDL also allow you to assign the numeric encodings (for example, in the later stages of a top-down design). A typical example is to specify the exact numeric code of the instructions add, Lda, Ldb.

Be aware that during logic synthesis an actual numeric coding is chosen, which might introduce invalid states. For example, in *Figure 4-1,* if 2 bits are used to encode type instruction:

   00 add
   01 Lda
   10 Ldb
   11 ---

In *Figure 4-1*, if instruct is a 2-bit register set to code 11, no case has been made for it. Therefore, it would be more prudent to actually specify a fourth value for type Instruction (e.g., invalid):

```
TYPE Instruction IS (add, Lda, Ldb, invalid);
```

Qualified Expressions

You cannot distinguish a between the character `'1'`, `bit '1'`, or `std_logic '1'`. When you see a '`1`' in quotemarks, you cannot be certain what type of literal it is, and so, you have an ambiguity of types. Therefore, you need *typecasting* to be explicit on the type of a value. You can *cast* its type in VHDL by using a *qualified expression*. For example:

```
bit'('1')
```

Another user example is:

```
type months is (April, May, June);
type name is (April, June, Judy);
```

`June` is of ambiguous type in some contexts. You can qualify `June` and explain exactly what `June` represents in VHDL with a qualified expression:

```
months'(June)
name'(June)
```

When a type has shared values with other types, you may need to clarify by casting a literal to a particular type.

The syntax is:

qualified_expression

> type' (literal or expression)

Do not confuse a qualified expression with a function that performs conversion.

Question: What is the difference between `integer(3.0)` and `integer' (3.0)`?

Answer:

Type Conversion

Because users and vendors create their own enumerated types, it is sometimes necessary to map one data type to another. Vendors may supply, or users may need to write, conversion functions when interfacing two designs that use different types. A variable or signal can be converted using a conversion function. The next example is a conversion example. Assume that you have two types: you have an incoming value of type **fourval**, and you want to convert it to an outgoing value named **value4**.

```
TYPE fourval IS ('X', 'L', 'H', 'Z');              Incoming Value
TYPE value4 IS ('X', '0', '1', 'Z');               Outgoing Value
FUNCTION convert4val (s: fourval) return value4 IS
BEGIN
   CASE s IS
      WHEN 'X' => RETURN 'X';
      WHEN 'L' => RETURN '0';
      WHEN 'H' => RETURN '1';
      WHEN 'Z' => RETURN 'Z';
   END CASE;
END convert4val;
```

Figure 4-2

The example uses the CASE statement on a parameter called **s**. **s** is of type **fourval**; the CASE statement breaks the values into four different cases **X**, **L**, **H**, and **z**. In each case, it returns a unique value. If you have one type and you want to convert it, you must write a function to do type conversion. Below is an example of a call to a conversion function:

```
Process...;
   Variable PDQ: fourval;
   Variable xyz: value4;
begin
   xyz := convert4val (PDQ);                       Function Call
end Process;
```

Figure 4-3

In *Figure 4-3*, the previously declared function is called from a process to convert data in **PDQ** into data in **xyz** of the corresponding type.

Question: In *Figure 4-2*, is RETURN **'X'** ambiguous type?
Answer:

Scalar Subtypes

In addition to the predefined types, you may find it convenient to define your own data types and subtypes. A subtype is based upon an existing type and is a restriction of that type in some way using a *range constraint* (see ***Figure 2-15***). For example, to declare a subtype `digit` of integers:

```
subtype digit is integer range 0 to 9;
```

Having defined this new type (`digit`), you can now declare variables or signals of this type. For example:

```
variable MSD, NSD: digit;
```

The subtype gives a convenient way of constraining an existing scalar type and using it as if it were a type. In this case, `digit` variables are restricted in assignment statements to be integer values in the range 0 to 9.

Note that this example is equivalent to, but more convenient than, specifying:

```
variable MSD, NSD: integer range 0 to 9;
```

Subtyping gives you type checking and a constraint on the original type. Other examples are:

```
type instr is (add, sub, mul, div, sta, stb, outa, xfr);
subtype arith is instr range add to div;          subset of instr
subtype pos is integer range 1 to 2147483647;     subset of integers
subtype nano is time range 0 ns to 1 us;          subset of time
```

Figure 4-4

When you use a variable assignment statement to assign a value, you must make sure that the value is of the right type and range.

When using a subtype, you should select the name of the type to be as meaningful as possible to other readers of your design. Type names are subject to the same syntax rules as identifiers.

Subtypes must be declared after the base type is declared. Subtypes must be declared before being used. A subtype uses a range constraint for a scalar type. Another use of a subtype is to create array types such as real and integer arrays (see Section 4.2). Subtypes are also used in resolution functions (see Section 9.2).

4.2 Composite Types - Arrays

Most programming languages provide array variables and records. Like them, VHDL has two composite types: *arrays* and *records*. Arrays have numerous elements of the same type and records have a group of objects of the same or different types. These are useful for describing busses, registers, or other arrays of hardware elements.

Arrays contain many elements of the same type. These elements can be scalar or composite. These elements or objects can be accessed by using an index. The only predefined array types are **bit_vector**, **string**, and **std_logic_vector**. You need to declare a new type for real arrays or integer arrays.

Index Range Declaration

Access depends upon declaration. For example:

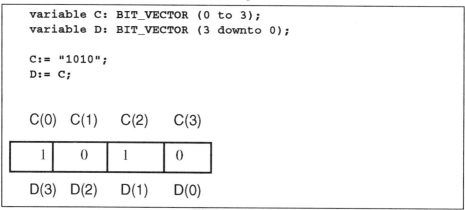

```
variable C: BIT_VECTOR (0 to 3);
variable D: BIT_VECTOR (3 downto 0);

C:= "1010";
D:= C;
```

C(0)	C(1)	C(2)	C(3)
1	0	1	0

```
D(3)  D(2)  D(1)  D(0)
```

Figure 4-5

The above example shows the declaration of **c** and **D** of type **BIT_VECTOR**. When you declare a bit vector object, which is an example of an array, you must declare it with an index constraint. In the example, **c** is declared to be an array of 4 bits from **0** to **3**. The indexes are **0** for the leftmost bit C(0) and **3** for the rightmost bit C(3). **D** is declared **3 downto 0** which is also a four-bit array. In this case, **3** is designated as the leftmost bit and **0** is the rightmost bit. Notice the reserved words **to** and **downto** in declaring the index range. VHDL has no particular standard for the ordering of bits or the numbering scheme. You can number from 1 to 4, or 4 to 7, etc. Having declared an array of particular range, it is important to use index references that are of the right type, in range, and in accordance with the original declaration.

NOT   **c(4)**   index out of range

              c(1.0)   wrong type

Array Assignment

VHDL Package STANDARD includes a bit type and BIT_VECTOR, which is an array of bits. Below is an example of a BIT_VECTOR assignment using some variables.

```
C := "1010";                     Constant BIT_VECTOR
C := S & T & M & W;              4 1-bit signals concatenated
C := ('1','0','1','0');          4 BIT Aggregate
C := 3;                          Invalid
```

Figure 4-6

c, which is a four-bit vector, gets the **BIT_VECTOR** assignment of a literal constant "**1010**". In the second example are four single variables, each of them one bit. They are concatenated together by the (ampersand) **&**. The example builds a four **BIT_VECTOR** on the right and assigns it to **c**, the four **BIT_VECTOR** on the left. The third example is an aggregate assignment. It requires parentheses and four individual bits, in single quotes, separated by commas. The example builds a four **BIT_VECTOR** *aggregate* and assigns it to **c**. Note that constant representation is not the same for **BIT** and **BIT_VECTOR** (single versus double quotes). The fourth case is an invalid assignment because **3** on the right is not a **BIT_VECTOR**.

Slice

A piece or *slice* of an array defines a subset of the array. The example below shows a valid assignment of a four **BIT_VECTOR**.

```
A: variable BIT_VECTOR (3 downto 0);
C: variable BIT_VECTOR (8 downto 1);

C(6 downto 3):=A;
```

Figure 4-7

A is a four **BIT_VECTOR** and **c** is an eight **BIT_VECTOR**. **C(6 downto 3)** designates the middle four bits of **c** and assigns it to four-bit **A**. Note that the length is the same on both sides of assignment. Any subrange or slice must declare subscripts in the same direction as initially declared:

```
c(6 downto 3)              NOT        C(3 to 6)
```

Aggregate

An array literal can contain a list of elements with both *positional* and *named* notation, forming a typed *aggregate*. The syntax is:

[*type_ name'*] ([choice =>] expression {, [others =>] expression})

type_name' can be any constrained array type. The optional choice specifies an element index, a sequence of indexes, or [others =>]. Each expression provides a value for the chosen elements and must evaluate to a value of the element's type. You can specify an element's index using either *positional* or *named* notation. Using positional notation, each element is given the value of its expression in order. For example:

```
variable X: BIT_VECTOR (1 to 4);
variable A, B: BIT;

X := BIT_VECTOR' ('1', A nand B, '1', A or B);
X := (1 => '1', 4 => A or B, 2 => A NAND B, 3 =>'1');
```

Figure 4-8

```
X(1)      gets '1'
X(2)      gets (A nand B)
X(3)      gets '1'
X(4)      gets (A or B)
```

Using named notation, [choice =>] specifies one or more elements of the array. [choice =>] can contain an expression (such as (I mod 2) =>) to indicate a single element index, or a range (such as 3 to 5 => or 7 downto 0 =>) to indicate a sequence of element indexes.

An aggregate can use both positional and named notation, but positional expressions must come before any named [choice =>] expressions. You do not need to specify all element indexes in an aggregate. All unspecified values are given a value by including [others =>] expression as the last element of the list. The next example shows an aggregate. It assigns a vector c, which is an eight bit BIT_VECTOR using B, which is of type bit.

```
variable B: BIT;
variable C: BIT_VECTOR (8 downto 1);

C := BIT_VECTOR' ('1', B, others => '0');
```

Figure 4-9

The example shows an aggregate assignment of eight bits on the right and eight bits on the left, where the eight bits on the right come from various sources. A qualified expression, designating a BIT_VECTOR , ensures that the type is correct.

A value with eight bits is in parentheses (starting with the leftmost bit, which is a '1', then B). The six remaining bits (6 downto 1) are assigned 0. The symbol => is read as "gets".

Aggregates can be used to express array literals, since they specify an array type and the value of each array element.

Variable Indexes

The next example uses a BIT_VECTOR with indexes inside of a loop.

```
.......
.......
BEGIN
PROCESS....
   variable info : BIT_VECTOR(0 to 49);
   variable start: INTEGER;
   variable byte_out: BIT_VECTOR(0 to 7);
BEGIN
   FOR i IN 0 to 7 LOOP
      byte_out(i) := info(i + start);
   END LOOP;
END PROCESS;
```

Array
Access

Figure 4.-10

info is a variable of 50 bits. byte_out is a BIT_VECTOR of eight bits, 0 to 7. The example finds a particular eight bits in the 50-BIT_VECTOR and copies eight successive bits and moves them to byte_out. The position number of the starting point is called start, which has an integer value. Notice the assignment statement, the index position starting at start + i where i is initialized to 0. During the first iteration, the bit is copied to byte_out(0). The loop runs eight times, adding 1 each time to i until i = 7. The last iteration copies a bit of data at start + 7 and moves it to byte_out(7). The example illustrates transporting bits using a loop and using indexes with different values for accessing different portions of two different vectors.

Also see Section 9.3 for length and range attributes.

Array Type Declaration

For integer or real arrays, you need to declare a new type. When you declare a new type that is an array, you need to declare:

> • The name of the type
>
> • The type of the elements
>
> • The number of index ranges
>
> • The type of the index
>
> • The range of index values (optional)

The syntax is:

type name **is array** [index_constraint] **of** element_type
    index_constraint :
        [range_spec]
         index_type **range** [range_spec]
         index_type **range** < >

Examples:

```
TYPE word8 IS ARRAY (1 to 8) OF bit;            ┐ Equivalent
TYPE word8 IS ARRAY (integer range 1 to 8) OF bit;  ┘ Statements
TYPE word IS ARRAY (integer range <>) OF bit;
TYPE RAM IS ARRAY (1 to 8, 1 to 10) OF bit;
```

Figure 4-11

The **<>** designates an unconstrained range, which is described later.

After a new type is declared as above, it can then be used in a signal or variable declaration. For example:

```
Variable mystuff: word8;                 8 bits
variable yourstuff: word (1 to 10);      10 bits
```

An enumerated type or a subtype also can be used to designate the range of index values. For example:

```
type instruction is (add, sub, Ida, Idb, sta, stb, outa, xfr);
subtype arithmetic is instruction range add to sta;
subtype digit is integer range 1 to 9;

type Ten_bit is array (digit) of Bit;
type Inst_flag is array (instruction) of digit;
```

Figure 4-12

Multi-Dimensional Arrays

Two-dimensional arrays can be useful for simulating ROMs and RAMs. VHDL allows multiple-dimension arrays. However, you must declare a new array type before you declare your variable or signal array. An array can have more than one index range; i.e., they can be two-dimensional. In the example below, type memory is declared to be a two-dimensional array of bits. The first index range is 0 to 7; the second is 0 to 3. The constant ROM is declared, of type memory (it literally is an array of bits). The bits are organized four per word, each word is in parentheses, and there are eight words separated by commas. You then refer to the array ROM with 2 indexes to get to a particular bit. data_bit receives ROM, word number 5, bit number 3 in the example reference.

```
TYPE memory IS ARRAY (0 TO 7, 0 TO 3) OF bit;
CONSTANT ROM: memory:=(      ('0','0','0','0'),
                             ('0','0','0','1'),
                             ('0','0','1','0'),
                             ('0','0','1','1'),
                             ('0','1','0','0'),
                             ('0','1','0','1'),
                             ('0','1','1','0'),
                             ('0','1','1','1'));


Example Reference:          data_bit := ROM (5,3);
```

Figure 4-13

Array of Arrays

It is also possible to have an array of arrays. The example declares TYPE word, which is an array of bits from 0 to 3 (words are four bits wide). memory is a type which is an array of 5 words (0 to 4). There is a constant array, rom_data, where each element itself is an array of type memory that contains the values shown. One level of subscripting is used to get to a particular word. A reference rom_data (addr) picks out a four-bit word and sends it to a variable of type word. rom_data (addr) (index) refers to a bit.

```
TYPE word IS ARRAY (0 TO 3) OF Bit;
TYPE memory IS ARRAY (0 TO 4) OF word;
VARIABLE addr, index: integer;
VARIABLE data: word;
CONSTANT rom_data: memory := (  ('0','0','0','0'),
                                ('0','0','0','1'),
                                ('0','0','1','0'),
                                ('0','1','1','0'),
                                ('0','1','1','1'));

data := rom_data (addr);


rom_data(addr)(index)              To access a single bit:
```

Figure 4-14

In VHDL you can have single-dimensional arrays, multiple-dimensional arrays, and arrays of arrays that can be arbitrarily complex. You must declare the array type before you declare a variable or signal array.

Unconstrained Array Type

Declaring array types with constant values for the ranges can be inconvenient. You might want to declare a type that has a variable number of elements to be determined when you actually declare a variable or signal of that type. An unconstrained array allows you to declare an array type without a particular range of values. For example, in the Package STANDARD, the unconstrained array type declaration is used to define **BIT_VECTOR**.

```
type BIT_VECTOR is array (natural range <>) of BIT;
```

Notice that the type of index is type **NATURAL**, which signifies non-negative integers. The range signified by **RANGE <>** (symbol called a "box"), means that the specific range will be filled in later when you use this type and its particular value. When you declare a BIT_VECTOR, signal, or variable, you must specify an index constraint. Also see Chapter 9 for information on length and range attributes.

Question: Can you declare a type for an unconstrained array of real numbers? If so, do it and name the type **real_array**.

Answer:

Array Index Constraint

The example below illustrates providing constraint in an index declaration:

```
VARIABLE xyz: BIT_VECTOR (1 TO 20);
```


Array Subtype

It may be convenient to declare a new type (subtype) of an existing array type. For example:

```
subtype byte is bit_vector (7 downto 0);
subtype twenty is bit_vector (1 to 20);
```

In this case, an index constraint is used on the unconstrained array type `bit_vector` to declare a new subtype. Variable and signed declarations can now use the subtype. For example:

```
Variable my_stuff: twenty;
```

The examples in *Figure 4-15* show some type definitions and associated signal declarations, followed by legal and illegal type conversions.

Legal Type Conversions

```
type INT_1 is range 0 to 10;
type INT_2 is range 0 to 20;

type ARRAY_1 is array (1 to 10) of INT_1;
type ARRAY_2 is array (11 to 20) of INT_2;

subtype MY_BIT_VECTOR is BIT_VECTOR(1 to 10);
type BIT_ARRAY_10 is array (11 to 20) of BIT;
type BIT_ARRAY_20 is array (0 to 20) of BIT;

signal S_INT:INT_1;
signal S_ARRAY:ARRAY_1;
signal S_BIT_VEC:MY_BIT_VECTOR;
signal S_BIT:BIT;
```

Integer Type Conversions

```
INT_2'(S_INT)
```

Similar Array Type Conversions

```
BIT_ARRAY_10' (S_BIT_VEC)
```

Illegal Type Conversions

`BOOLEAN (S_BIT);`	Cannot convert between enumerated types
`INT_1' (S_BIT);`	Cannot convert enumerated types to other types
`BIT_ARRAY_20' (S_BIT_VEC);`	Array lengths not equal
`ARRAY_1' (S_BIT_VEC);`	Element types are not convertible

Figure 4-15

4.3 Composite Types - Records

Records group objects of different types into a single object. These elements can
be of *scalar* or *composite* types and are accessed by name. Each field of the record
can be referenced by name (selected name). Records provide abstract modeling
capability.

```
TYPE two_digit IS
   RECORD sign:  bit;
            msd:  integer range 0 to 9;
            lsd:    integer range 0 to 9;
END RECORD;


PROCESS
VARIABLE ACNTR, BCNTR: two_digit;
 BEGIN
   ACNTR.sign := '1';
   ACNTR.msd := 1;
   BCNTR.lsd := ACNTR.msd;
   BCNTR   := two_digit'('0',3,6);
END PROCESS;
```

Figure 4-16

The example illustrates defining a new type **two_digit** using a record to represent
a sign-magnitude value containing two digits (i.e., from -99 to +99). Two counter
variables, **ACNTR** and **BCNTR**, are declared of this record type. Individual elements
of the variable are accessed using the element name appended (**ACNTR.sign**) as
originally declared in the record. The last example illustrates an aggregate
assignment to the three fields of **BCNTR**. Named aggregates are allowed using
field names as indexes. You also can have an array of records.

4.4 Alias Declaration

An alias is an alternate name assigned to part of an object, which allows simple
access. For example, suppose a 9-bit bus **count** has three elements: a sign, the
msd, and lsd. You could operate on each named element using an alias
declaration, as shown below.

```
signal count: bit_vector(1 to 9);
   alias sign: bit is count(1);
   alias  msd: bit_vector(1 to 4) is count(2 to 5);
   alias  lsd: bit_vector(1 to 4) is count(6 to 9);
```

Figure 4-17

You can then access any element of **count** by name:

```
sign := '1';
msd  := "1001";
count:= "1_1001_0000";
lsd  := msd;
```

4.5 Predefined Types: Text and Lines

Two other data types that are predefined in VHDL are type *text* and type *line*. Text and lines are used for input and output operations during simulation. These types are used in a `file` declaration with read, write, and end-of-file functions. Text files and lines occur in processes and subprograms. They allow reading and writing of ASCII text files. Both provide formatted input and output. Files of type text are treated as groups of lines.

In the library STD, there is a set of functions and types in Package TEXTIO that are predefined. These functions allow you bring in and send out data between files and variables. The input procedures in the Package TEXTIO are `readline`, which reads a line of text out of a file, and `read`, which reads an item off a particular line. The output procedures in the TEXTIO package are `writeline` and `write`. For example:

```
readline (F: In TEXT; L: out LINE);      Reads a line from the file F
                                            into variable L

read (L:Inout Line; ITEM:integer);       Reads an item out of line L into
                                            variable ITEM
```

Figure 4-18

Reading data from a file is a two-stage operation. For example:

```
a: process ...
file testvectors: TEXT is in "test.vec";
variable L : Line;
variable av, bv : bit_vector (3 downto 0);
begin
    readline (testvectors, L);
    read (L,av);
    bv:=av;
```

Figure 4-19

In the example above, you have a process **a** and are identifying a logical file named `testvectors` of type `text` that is called `test.vec` in the file system. There is a variable `L` of type `Line` and four-bit variables `av` and `bv`. The operation after the `begin` reads a line from the `testvectors` file into the variable `L`. The subsequent read inputs the first four-bit value of line `L` and puts it into variable `av`.

You need to create a reference to this library package before you do text I/O:

```
use STD.TEXTIO.ALL;
```

In addition, there are two functions:

```
ENDFILE (filename)
ENDLINE (linename)
```

which return Boolean values that indicate an end-of-file or end-of-line condition. Below is a complete example using this package.

```
USE STD.TEXTIO.ALL;
ENTITY copy4 IS
END copy4;
ARCHITECTURE first OF copy4 IS
BEGIN
PROCESS(go)
FILE instuff: TEXT IS IN "/path/test.data";
FILE outfile: TEXT IS OUT "/path/new.data";
VARIABLE L1, L2:LINE;
VARIABLE av:bit_vector(3 downto 0);
BEGIN
WHILE NOT (ENDFILE(instuff)) LOOP
   READLINE (instuff, L1);          Reads a line from the input file
   READ (L1, av);                   Reads a value from the line
   WRITE (L2, av);
   WRITELINE (outfile, L2);
END LOOP;
END PROCESS;
END first;
```

Figure 4-20

Datafile `test.data` contains

```
0011
00_11
1100
16#F#
1010   1011
```

Underscores are allowed and ignored. # signs indicate a number in a different radix (for example, base 16). Data of type `bit` and `bit_vector` do not have quote marks. Enumerated types, which utilize identifiers (see Section 4.1), cannot be input or output with TEXTIO.

4.6 Access Types

Access types are not supported by synthesis but can be use during simulation. Access types are similar to pointers in programming languages. They assist in building up structures in simulator memory, which are of varying sizes. They permit the user to dynamically declare when memory is needed for an item and to allocate memory from the simulator as needed to hold newly created items. Variables of type access are pointers which can be manipulated. A typical use of access types is to create a linked list structure.

One operation is the NEW operator, which causes the simulator system to allocate storage for an item (typically a record) and it returns the *address* of where the new item is stored in simulator memory. Another operation is DEALLOCATE, which deletes an item from memory. In the examples below:

```
type two_digit is record ...       -- see Figure 4-16
type pntr is access two_digit;     -- pntr is a pointer, access type
variable xptr: pntr;               -- a pointer variable to access a  record
```

Example of storage allocation:

```
xptr := new two_digit;             --allocates space for a record
                                     address is in xptr
```

An example of storage deallocation:

```
deallocate (xptr);
```

To access elements in the record, use selected field name with the pointer. For example:

```
xptr.msd := "1001";
```

Type Declaration

Type declarations must occur before a type is used. They can be declared as diagrammed below. A type declared in an entity cannot be used in the port specification of that entity because the port declaration occurs prior to type declarations. Type declarations should be declared in a package.

type_declaration

```
type name is definition;
type name;*
```

Entity

```
entity E is
type_declaration

begin..............


end E;
```

Process

```
A: process
type_declaration

begin..............


end process A;
```

Package Declaration

```
package A is
type_declaration


end A;
```

Block

```
A: block
type_declaration

begin..............

end block A;
```

Procedure

```
procedure A (...) is
type_declaration

begin..............

end A;
```

Package Body

```
package body A is
type_declaration


end A;
```

Architecture

```
architecture A of E is
type_declaration

begin................


end A;
```

- **Summary**

1. *Enumerated* types allow user-defined types. Enumerated types can be *literal* values or *identifiers*. (Section 4.1)

2. A *qualified expression* can cast the type of a literal. (Section 4.1)

3. *Subtypes* constrain a range or subset of an existing type. (Section 4.1)

4. *Composite* types consist of arrays and records. (Sections 4.2)

5. *Arrays* are collections of elements or objects of the same type. These elements can be indexed. You can also define slices of an array. Arrays can be single-dimensional or multi-dimensional. (Section 4.2)

6. *Records* are groups of objects of the same or different types. Each field of the record can be referred to by name. (Section 4.3)

7. *Files* and *line* types are used for input, output, and end-of-file functions. (Section 4.4)

8. *Types* can be declared in an entity, package, architecture, block, process, or procedure. They cannot be declared in a configuration.

-   Questions and Answers

Question: What is the difference between `integer (3.0)` and `integer' (3.0)`?

Answer: `integer(3.0)` is a legal call to the built-in conversion function from real to integer.
`integer'(3.0)` is an invalid qualified expression. You cannot cast the type from real to integer.

Question: In *Figure 4-2*, is `RETURN 'x'` an ambiguous type?
Answer: No, the return value is declared to be of type `value4`.

Question: Can you declare a type for an unconstrained array of real numbers? If so, do it and name the type `real_array`.
Answer: `type real_array is array(integer range <>) of real;`

- **Lab #4**

Using TEXTIO Package

Rewrite the temperature process in Lab # 2 to read a value from a file. Create a
`text/io` file using a text editor and try to process it with values.
0.0
40.0
22.0
18.0

5

Signals & Signal Assignments

This chapter discusses the use of signals for component interconnection and process communication. It contains the following sections:

- •Structural Netlisting
- •Process Communication
- •Signal Declaration
- •Entity Signal Port Declarations
- •Signal Assignment in a Process
- •Signal Delay
- •Sequential Signal Assignment
- •Simulation Cycle
- •Simulation and WAIT
- •Sensitivity List

Signals provide communication between processes and between components. Unlike local variables that are declared in a process, signals provide global communication within a design.

Signals are used to wire structural designs (see Section 5.1 and Chapter 7). Signals are also used to communicate between processes (see Section 5.2). Within a process, signal assignments can send data to other processes. Signal assignments are scheduled in simulated time and are not treated like variable assignments. (However, signal values can be used on the right side of variable assignment statements.) Variables are used as in most other programming languages. Signals have unusual characteristics that are very "unprogram like".

VHDL '92 supports global variables declared in an architecture and shared among multiple processes.

5.1 Structural Signals in Netlisting

Design entities usually contain instances of previously compiled design entities (e.g.,components). The top level design is frequently done in a structural style with library units *instanced* as components "netlisted" together using signals. Netlists are common in most electronic-design automation systems today and are a familiar concept to engineers. Programmers may find this notion unfamiliar. (Also see Section 1.3 and Section 5.4. and Chapter 7.)

Signals declared in the port declaration of an entity describe a component's external input/output connectors. The example shows COMPARE with signals A, B, and C of type bit declared in the entity. Two of the signals A, B are inputs and one of them, C, is an output. Only these three signals are accessable (visible) externally. Additionally, they can be used internally in the architecture.

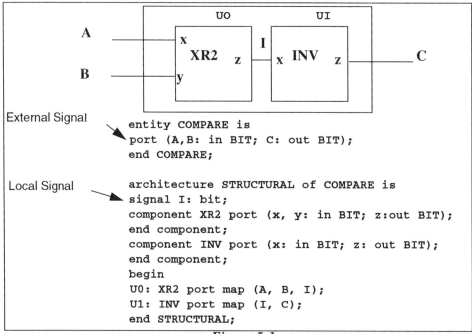

```
External Signal          entity COMPARE is
                         port (A,B: in BIT; C: out BIT);
                         end COMPARE;

Local Signal             architecture STRUCTURAL of COMPARE is
                         signal I: bit;
                         component XR2 port (x, y: in BIT; z:out BIT);
                         end component;
                         component INV port (x: in BIT; z: out BIT);
                         end component;
                         begin
                         U0: XR2 port map (A, B, I);
                         U1: INV port map (I, C);
                         end STRUCTURAL;
```

Figure 5-1

In addition to the three external signals, the architecture STRUCTURAL of the design entity COMPARE includes a local declaration of signal I, used for connecting two components, an XR2 gate and an inverter INV. This signal I cannot be accessed externally and has <u>no</u> direction (in or out).

The two components are declared in a component declaration before they are used. The signal ports x, y, z in these declarations are connectors to the XR2 and INV gates; these names are identical to the original entity port declarations of the XR2 and INV entities. The instance U0 uses actual entity signals: A to connect to port x of XR2, B to connect to input port y of XR2, and local signal I to connect to output port z of XR2. In instance U1, I connects to port x of INV and the entity signal C connects to output port z of INV. This example illustrates two kinds of signals --the internal signal I connects component U0 to U1; the external signals are A, B, and C.

5.2 Process Communication

Signals can also be used for activating or synchronizing processes inside an architecture; one process can be waiting for a signal from another process. Below is a diagram illustrating two processes, where process **A** is waiting until process **B** is done.

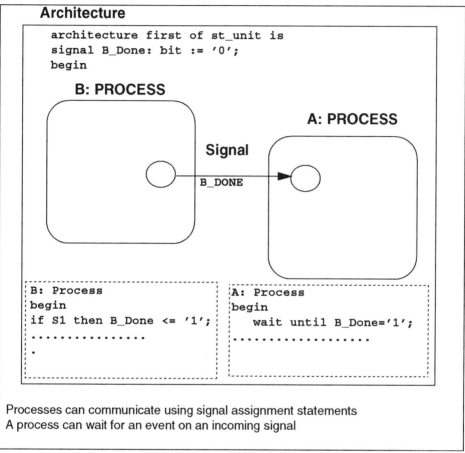

Architecture

```
architecture first of st_unit is
signal B_Done: bit := '0';
begin
```

B: PROCESS

A: PROCESS

Signal

B_DONE

```
B: Process
begin
if S1 then B_Done <= '1';
................
    .
```

```
A: Process
begin
   wait until B_Done='1';
...................
```

Processes can communicate using signal assignment statements
A process can wait for an event on an incoming signal

Figure 5-2

Process **A** includes a `wait until  B_Done = '1'`, which suspends execution of process **A**. The signal `B_Done` is assigned in process **B** and, when set to `'1'`, activates process **A**. Typically, a process has a `wait` for an event on an incoming signal. The event must be a change (for example, '0' becomes a '1' or '1' becomes a '0'). This example illustrates process synchronization using a single signal. Since both Process **A** and **B** are in the same architecture, signal **B_DONE** is declared within the architecture and is available and "visible" to all its processes.

Note: Variables declared in process B cannot send values to another process. Signals can be used for interprocess communication. In VHDL '92 global variables can also be used.

5.3 Process/Component Connection - Testbench

Signals can be used to connect components (Section 5.1) , processes (Section 5.2), and connect process to components. *Figure 5-3* illustrates this style of design in which a process is used as a testbench for exercising component UUT, which is a compiled VHDL design. Rather than having a non-VHDL test, users often code their *functional tests* as a VHDL process to validate a VHDL *design*. These tests are run on the simulated design in a VHDL simulator.

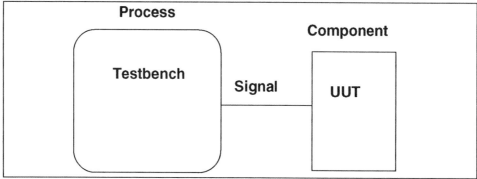

Figure 5-3

The form of the testbench may be one of the following styles.

Ad Hoc

During debugging, a process is written by the designer, which is an ad hoc collection of stimulus cases to exercise the basic functions of the design unit under test (UUT).

Algorithmic

Another way to generate a stimulus for exercising a component is to write a simple algorithm. An example is a loop that increments a signal value, counting over a complete range (e.g., for testing a decoder or arithmetic unit). You need to observe and check the results of the simulator output.

Vector File

A more structured approach is to use a process to read a file of test vectors from a data file. In this case, the user formats the vectors as appropriate for the design. Each vector is read and used as a stimulus signal for the UUT. The expected response is in the file; the process monitors whether or not the behavior of the UUT matches the expected outcome designated in the file of test vectors.

These tests might be the basis of a physical device test to test an ASIC's functional behavior. Another use is for regression testing, to prove that a modification of a design of a UUT still maintains the original design specification.

5.4 Signal Declaration

Signals can be used to communicate between processes or components. A signal must be declared before it can be used (see *Figure 5-2*); B_Done is declared as a signal in the architecture first prior to its use in the processes. Although a signal can be used in a process, it cannot be declared in a process. Signals can be declared in a number of places. Global signals can be declared in packages. If declared in an entity or port declaration, signals can be used in any architecture of that entity. Declared signals within a particular architecture are local to that architecture. Signals are initialized with an := just like variables; *signal assignments* are written with a special symbol <=.

Signals can be declared:

- In packages - global signals

- In entity declaration sections - entity global signals

    Ports: in, out, inout, buffer
    Other declarations
- In architecture declaration sections - architecture global signals

Figure 5-4 shows four kinds of signal declarations. Signals vcc and ground are global signals in a package called sigdecl. Entity signals data_in and data_out are declared as input/output ports; their types are bit. In addition, there is a signal that is not an I/O port, but is global to all architectures within the entity called sys_clk. Any architecture of this particular entity board_design can use these signals. In the architecture called data_flow for this entity, there is a signal called int_bus. Within this architecture, any process or any component can use this signal. The example shows the four different types of signal declarations, all which must occur before a signal is used within an architecture.

```
PACKAGE sigdecl IS
   SIGNAL vcc: bit := '1';      ◄──────── Global Signals
   SIGNAL ground : bit := '0';
END sigdecl;

ENTITY board_design IS
   PORT (data_in: IN bit;       ◄────────┐ Entity Global Signal
   data_out: OUT bit);                   │
   SIGNAL sys_clk : bit := '1'; ◄────────┘
END board_design;

ARCHITECTURE data_flow OF board_design IS
   SIGNAL int_bus: bit;         ◄──────── Architecture Local Signal
BEGIN
```

Figure 5-4

5.5 Entity Signal Port Declarations

When you declare a signal port of an entity, you declare its name, direction, and type, and optionally a default initial value. The syntax is:

port_declaration

```
port (names : direction type [:= expression] [;more_ports]);
```

The default direction is *input*, but it is important to declare otherwise (see **Figure 5-5**). A `buffer` is the same as `inout` except there can only be one driver or source. The box below outlines appropriate entity signal directions and their usage.

Signal Direction	Usage
In	Right side of variable or signal assignment
Out	Left side of signal assignment
Inout	Both above
Buffer	Both above (1 driver only)

Figure 5-5

The example below shows the name, direction, and type of two entity port signals:

```
PORT (data_in: IN bit; data_out: OUT bit);
```

Question: In **Figure 5-6** below, what is the likely signal direction for `data_out`?

Answer: _____

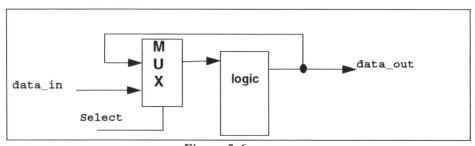

Figure 5-6

Type declarations must occur before a type is used in a port. A type declared in an entity cannot be used in the port specification of that entity because the port declaration occurs prior to type declarations (see Section 1.3). Type declarations should be declared in a previously compiled package.

In the example below, `MyType` must be previously declared in `MyPkg`.

```
port (B, A: in MyLib.MyPkg.MyType);
```

A port can be given a default initial value [:= expression].

5.6 Signal Assignment in a Process

Signal assignments *schedule* a value at some instant in simulated time to a target previously declared as a signal object. Such an assignment defines a *driver* of a signal. Signal assignments are *sequential* within a process or *concurrent* outside of a process (see Chapter 6). Within a process, signal assignment scheduling is delayed until a simulation cycle is run, triggered by the execution of a WAIT statement.

```
Process
BEGIN
sys_clk <= NOT (sys_clk) AFTER 50 NS;
int_bus <= data_in after 10 ns;
data_out <=my_function (int_bus) after 10 ns;
wait ....
END;
```

Figure 5-7

The example in *Figure 5-7* illustrates a series of sequential signal assignment statements. `sys_clk` gets the complement of the current value after `50 NS`. `int_bus` gets the same value currently as `data_in`. `data_out` gets assigned a value by calling a function using the current value of `int_bus`. The assignments are scheduled only after the WAIT statement is executed by the simulator.

Signal assignment statements can schedule a number of values at different time points. This capability is useful for describing clocks and other repetitive signal waveforms. For example:

```
S <= '1' after 4 ns, '0' after 7 ns;
T <= 1  after 1 ns,  3 after  2 ns,  6 after  8 ns;
```

Figure 5-8

Note: Within a process a signal should have only one single driver at a time. In the example below, only the second assignment to `xyz` is used.

```
Process
begin
   xyz <= 1 after 4 ns;
   PDQ <= 10 after 5 ns;
   xyz <= 2 after 4 ns;
wait ...

Only the second assignment to xyz  is effective
```

Figure 5-9

5.7 Signal Delay

When a signal assignment is simulated, a signal change occurs at a precise simulated time. Signal delays are often used in modeling the abstract behavior of a physical system or in very detailed models with complex formulas for expressing the delay.

```
        signal x, y: integer;        It may not be possible to build
        .....                        hardware to this specification.
        process
        begin                        This updates x after 10 ns
        WAIT ON Y;                   (not 9.9999 or 10.0001).

    x <= y + 1 after 10 ns;
```

Figure 5-10

In this example, **x** gets the value **y+1** after 10 nanoseconds from a change in **y**. You are giving instructions for **x** to be updated exactly 10 nanoseconds later. This statement does not mean 9.9999 or 10.00001, but exactly 10 nanoseconds. Unfortunately, real hardware may not be able to do exactly what VHDL signal assignments specify. Therefore, it may be possible to simulate them, but it may not be possible to build them exactly.

As described in Section 5.6, signal assignment statements define a precise simulation time for driving a value to a signal. This feature is useful in more abstract specifications of hardware, but it is probably not meaningful for logic-synthesis tools.

The next example shows a logic loop. Signal **x** is feeding back to an adder; the logic loop is reconciled (in terms of simulator behavior) because the new output value is delayed by 10 nanoseconds from the inputs. It is difficult to build a combinational-logic circuit that works like this; generally, you get a race condition or oscillation. In terms of simulation, the code is correct because you are describing the abstract behavior of a hardware element. In logic synthesis, this type of description requires refinement by the designer (for example, to introduce a clocked register for building a counter).

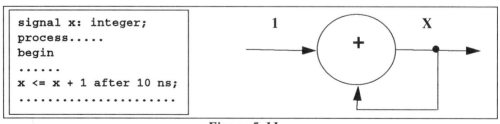

Figure 5-11

Zero Delay

Although first-generation simulation systems have difficulty ordering events with zero delay, VHDL simulation semantics define an unambiguous behavior. Signal assignments can specify zero delay by not providing an AFTER clause. The next example shows a multiplexor using a signal assignment with zero delay. This behavior is an abstraction of a real multiplexor circuit, which would have non-zero delay.

```
entity VAR is
port (A     : in BIT_VECTOR (0 to 7);
      INDEX : in INTEGER range 0 to 7;
      OUTPUT : out BIT);
end VAR;

architecture VHDL_1 of VAR is
begin
   process
   begin
     OUTPUT <= A(INDEX);          Scheduled in 0 ns
     wait .....;
.........
end VHDL_1;
```

Figure 5-12

The entity **VAR** has an 8-bit input port **A**. One of these single bits is assigned to **OUTPUT**. Bit selection is done using **INDEX**, which is an **INTEGER** in the range **0** to **7**. The entire architecture for the multiplexor is one statement; **output** gets **A(index)**. Because this example is a signal assignment, without an AFTER clause, the assignment is scheduled to happen after 0 nanoseconds; 0 nanoseconds has a precise meaning in VHDL simulation. However, even though there is no delay, the signal assignment statement is not performed until the WAIT statement is executed .

When a signal assignment statement occurs within a process, the actions proscribed do not take effect (scheduled) until a simulation cycle occurs (e.g., WAIT executed, shown in *Figure 5-13*). This behavior is considerably different than *variable assignment statement* execution in a process, which occurs immediately and does not depend upon a WAIT instruction.

The main differences between signal assignment and variable assignment are:

Signal Assignment	Variable Assignment

Signal values are scheduled
Signal assignments can have a delay
Signal values are updated
  only after wait is executed
  (e.g, simulation cycle)
	Variable values are not scheduled
	Variable assignment are specified without a delay
	Variable values are updated immediately

Examples:

```
x <= 1;            0 delay
Wait ...;          Assignment takes place
                   after the WAIT is executed

x <= y;            0 delay
y <= x;
Wait ...;          Both assignments,
                   exchange occurs after WAIT is executed

v := 1;            Variable assignment occurs immediately
s <= v;
a := s;            a gets old value of s,
WAIT ...;          v gets 1,
                   s gets v (e.g. 1) after the WAIT is executed

x <= 1;
x <= 2;
Wait for 0 ns;     After WAIT is executed, x gets value 2
```

Figure 5-13

5.8 Sequential Signal Assignment Hazards

The use of signal assignments inside of a process can cause invalid or incorrect results. This case is shown by the incorrect mux example **wrong**.

```
ENTITY mux IS
PORT (x, y: IN BIT;
      select: in Boolean;
      p: OUT BIT);
END mux;

ARCHITECTURE wrong OF mux IS
SIGNAL muxval: INTEGER range 0 to 1;
BEGIN
PROCESS
BEGIN
   muxval <= 0;
   IF (select) THEN
       muxval <= muxval + 1;
   END IF;

   CASE muxval  IS
       WHEN 0 => p <= x AFTER 10 NS;
       WHEN 1 => p <= y AFTER 10 NS;
       END CASE;
       wait on x, y, select;
END PROCESS;
END WRONG;
```

Figure 5-14

Look through the process, notice that it first initializes **muxval** to **0** . The entity port boolean signal **select** controls whether **x** or **y** goes to the output signal **p**. Although **muxval** appears to be initialized to **0**, it can take on the values **0** or **1**, depending upon the value of the **THEN** clause, which is controlled by **select**. It seems perfectly appropriate then, to follow the assignments to **muxval** with a CASE statement. The CASE uses the values of **muxval** (being **0** or **1**) and assigns to **p** the value of one of the inputs after **10 NS**.

The example looks like a perfectly accurate description of a 2-to-1 multiplexor. Notice, however, that the example uses a signal assignment to **muxval**. The problem is that the testing of **muxval** in the CASE statement does not test the value currently being assigned. The expression **muxval <= 1** is a scheduled event and it does not actually occur until the WAIT statement is executed. Using **muxval** immediately after it has been assigned means to use its prior value. This meaning may not be what the designer intends! (The signal **select** could have been tested in the CASE statement directly instead of using **muxval**.)

You can fix the hazard by avoiding the use of signal assignment and by using variable assignments instead.

```
ENTITY mux IS
PORT (x, y: IN BIT;
      select: in Boolean;
      p: OUT BIT);
END mux;

ARCHITECTURE better OF mux IS
BEGIN
PROCESS
  VARIABLE muxval : INTEGER range 0 to 1;
BEGIN
  muxval := 0;                                 Note variable
  IF (select) THEN
      muxval := muxval + 1;
  END IF;

  CASE muxval  IS
      WHEN 0 => p <= x AFTER 10 NS;
      WHEN 1 => p <= y AFTER 10 NS;
      END CASE;
      wait on x, y, select;
END PROCESS;
END better;
```

Figure 5-15

In the *Figure 5-15*, the code has been changed to use **VARIABLE muxval**. Variable assignments take place immediately. Therefore, by the time the simulator executes the CASE statement, **muxval** has the correct *current* (not previous) value.

The examples in *Figures 5-13* show the difference between variable assignment and signal assignment. You can see that the use of signal assignments in a process can be error prone; variable assignments are less likely to confuse.

5.9 Simulation Cycle

First-generation simulators used a technique CAD developers call a *one-list algorithm*, which is relatively fast but cannot handle parallel zero delay events such as exchanging **a** and **b**. The meaning and problems that are associated with this zero delay signal assignment can be illustrated by a simple example:

```
A  <=  B;                    0 delay
B  <=  A;                    0 delay
```

This example would not exchange the values of **a** and B, but would give both **a** and **b** the old value of **b**, using a one-list algorithm.

VHDL uses a *two-list algorithm*, which tracks the previous and new values of signals. In this method, expressions are first evaluated, then signals are assigned new values. In VHDL, the example code performs a data exchange between the two signals **a** and **b** at some point in simulated time. In operation, the old values of **b** and **a** are fetched and scheduled for assignment, for zero delay, after a subsequent WAIT statement is executed.

The ordering of zero delay events is handled with a ficticious unit called *delta time*. Delta time respresents the execution of a simulation cycle without advancing simulation time (see Figure 5-16 for further explanation).

The key points of simulation and delta time are:

- The simulator models zero-delay events using delta time.

- Events scheduled at the same time are simulated in specific order during a delta time step.

- Related logic is then resimulated to propagate the effects for another delta time step.

- Delta time steps continue until there is no activity for the same instant of simulated time.

The VHDL simulator cycle is shown in *Figure 5-16* below.

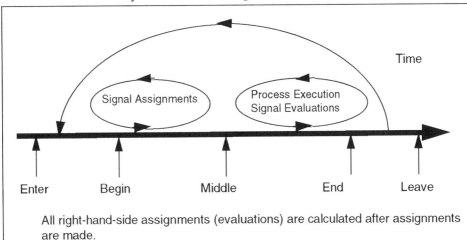

Figure 5-16

The execution of a WAIT statement triggers a simulation cycle to be entered. The simulation cycle begins at Enter time. VHDL simulators do all signal assignments with previously computed and scheduled values. Since some signal values change, some processes may be activated. Then, the simulation cycle goes through the middle time and active processes are executed. Variable assignment statements are executed; expressions on the right-hand-side of signal assignments are evaluated. Elements that need to get scheduled are evaluated to get the appropriate values.

It may be necessary to return to Begin without actually advancing simulated time. This cycle is called a *delta time step*. Delta time is the iteration without the advancement of simulated time. A simulator could cycle around this loop several times doing evaluations and assignments without advancing simulated time. When no more processes are triggered, the simulation cycle goes to the Leave time, and simulation time is then advanced. The question below asks how signal assignments work inside a process.

```
Process
begin
    x <= 1;
    x <= 2;
    a <= x;
    x <= 3;
wait for 0 ns;
```

Figure 5-17

Question: What are the values of **x** and **a** after 1 simulation of one delta step?

Answer:

5.10 Simulation and WAIT Statements

A process can have signal assignments and variable assignments. There is a difference between these assignments. Signal assignments use scheduled time and variable assignments occur immediately. A process contains only sequential statements and normally has one or more WAIT statements in it. Signal assignments in a process do not take effect until a WAIT statement occurs. A WAIT statement suspends sequential statement execution, but permits signals that have scheduled values to be assigned.

There are four forms of WAIT. A WAIT ON allows us to suspend execution until there is an event on a signal(s). WAIT UNTIL, occurs with a conditional event. WAIT FOR waits a specific amount of time. WAIT waits forever.

Process WAIT Statement Examples

```
• WAIT ON a, b;
    Suspends execution until an event occurs on either a or b
• WAIT UNTIL a >10;
    Suspends execution of this process until an event occurs on a, and a >10
•  WAIT FOR 10 NS;
    Suspends execution of this process for 10 ns
•  WAIT;
    Suspends execution of this process forever
```

Figure 5-18

Note: There is no way to delete or kill a process in VHDL. Therefore, the unconditional form of the WAIT is used to suspend (i.e., kill) a running process. This form is convenient for processes that are used only for initialization.

5.11 Sensitivity List

Rather than having one or more WAIT statements inside a process, you can have a sensitivity list. A sensitivity list is a list of names for which a process is going to wait. A sensitivity list follows the word **PROCESS** in parentheses as shown in *Figure 5-19*.

A sensitivity list is equivalent to having a WAIT at the *end* of the process, *not* at the beginning of the process. You can have a sensitivity list or you can have WAIT statements but not both. In the example below there is a process waiting for a signal called **clk**. This code is equivalent to having a WAIT statement at the end of the process.

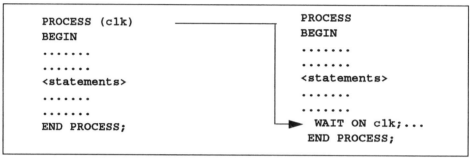

Figure 5-19

Question: What is the difference in behavior below?

Answer:

```
process (A,B)
begin
S <= A;
T <= B;
V <=S or T;
end process;
```

```
process (A, B, S, T)
begin
S <= A;
T <= B;
V <= S or T;
end process;
```

Note: All processes are run during startup until a WAIT is executed. Therefore, this process first runs during initialization.

- **Summary**

1. *Signals* provide communication between *processes* and between *design entities.* Unlike *local variables* that are local to a process, signals provide global communication within a design. Signals are also used for structural netlisting. (Section 5.0)

2. Signals declared in the port declaration of an entity describe a component's input/output connectors. You must declare a signal in an entity port or in a signal declaration statement before using it. (Section 5.1)

3. *Signal values* are scheduled in simulated time and are activated in a process when a WAIT is executed. *Variable values* are not scheduled and happen immediately. (Section 5.9)

4. Rather than having one or more WAIT statements inside a process, you can have a sensitivity list. A sensitivity list is a list of names for which a process is going to wait. A sensitivity list follows the word **PROCESS**. (Section 5.10)

5. A sensitivity list is equivalent to having a WAIT ON at the *end* of the process, *not* at the beginning of the process. *WAIT statements* can be used to dynamically control process suspension and execution. (Section 5.10)

- **Answers to Questions**

Question: In *Figure 5-6* below, what is the likely signal direction for `data_out`?
Answer: Buffer

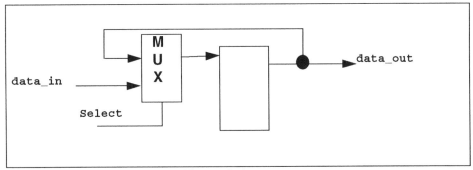

Figure 5-6

Question: What are the values of `x` and `a` after one delta step?
Answer: `x` is 3 and `a` is prior (old) value of `x`.

Question: What is the difference in behavior?
Answer: The process is activated twice - first because `A` or `B`
changed, secondly because `s` or `T` changed.

• Single-Pass Simulation
`process (A,B)`
`begin`
`S <= A;`
`T <= B;`
`V <=S or T;`
`end process;`
 	
• Will use old value of `s`, `T` for `v`.

• Two-Pass Simulation
`process (A, B, S, T)`
`begin`
`S <= A;`
`T <= B;`
`V <= S or T;`
`end process;`
 	
• Updates value of `s`, `T` for `v` in two delta time steps.

- **Exercise 1**

Write a 4-stage moving average process called P and place it in an architecture called moving4, using type real for input port NEWS and output port AVG. In your process, make sure to declare a new type for the array of real numbers (real_vect). This type can be limited to subscripts 1 to 4 or can be unconstrained. Declare a variable array named s for the 4 real values being remembered. Use a WAIT statement to get a new result, each 1 ns.

1. Create the definition for an input real sample (named NEWS).

2. Create the definition of 4 local storage signals (array named s), subscripts 1 to 4.

3. Write the code for shifting in a new value to s1, and s1 to s2, s2 to s3, s3 to s4.

Write the code for the Average calculation (named AVG) of four values in s.

- **Answer to Exercise 1**

Write a 4-stage moving average process called P and place it in an
architecture called moving4, using type real for input port NEWS and
output port AVG. In your process, make sure to declare a new type for
the array of real numbers (real_vect). This type can be limited to
subscripts 1 to 4 or can be unconstrained. Declare a variable array named
S for the 4 real values being remembered. Use a WAIT statement to get a
new result, each 1 ns.
1. Create the definition for an input real sample (named NEWS).
2. Create the definition of 4 local storage signals (array named S),
subscripts 1 to 4.
3. Write the code for shifting in a new value to S1, and S1 to S2, S2 to S3,
S3 to S4.
Write the code for the Average calculation (named AVG) of four values in
S.
Write the code for defining the output of the Avg

```
entity moving4 is
port (NEWS in : real; Avg out: real);
end moving4;

architecture first of moving4 is
   type real_vect is array (1 to 4) of real;
   process
   variable S: real_vect;
   Begin
   S(4):= S(3);
   S(3):= S(2);
   S(2):= S(1);
   S(1):= NEWS;
   Avg <= (S(1) + S(2) + S(3) + S(4))/4.0;
wait for 1 ns;
end process;
end first;
```

Can you rewrite the four assignments in just 1 statement using an
aggregate?

- **Exercise 2**

The FOR Loop

A LOOP statement includes a sequence of statements that can replicate logic, zero or more times. Expand the FOR loop to show the equivalent four statements.

```
architecture SHORTER of DECODER is
begin
process (BUS_A) -- note of type integer
begin

  for K in 0 to 3 loop
     D(K) <= (BUS_A = K);
  end loop;


end process;
end SHORTER;
```

-   **Answer to Exercise 2**

The FOR Loop

A LOOP statement includes a sequence of statements that can replicate logic, zero or more times. Expand the FOR loop to show the equivalent four statements.

```
architecture SHORTER of DECODER is
begin
process (BUS_A)
begin


    D(0) <= (BUS_A = 0);
    D(1) <= (BUS_A = 1);
    D(2) <= (BUS_A = 2);
    D(3) <= (BUS_A = 3);


end process;
end SHORTER;

```

8-Bit Odd Parity Checker

What is wrong with this code?

```
  procedure ODD_PARITY (A: In BIT_VECTOR (0 to 7); RESULT :
out BIT) is
begin
RESULT <=0;
for I in 0 to 7 loop
   RESULT <= RESULT xor A(I);
end loop;
end ODD_PARITY;
```

- **Answer to Exercise 3**

8-Bit Odd Parity Checker

What is wrong with this code?

```
procedure ODD_PARITY (A: In BIT_VECTOR (0 to 7);
                                RESULT : inout BIT) is
begin
RESULT <='0';
for 1 in 0 to 7 loop
   RESULT <= RESULT xor A(1);
end loop;
end ODD_PARITY;
```

The code has multiple signal assignments at the same
simulation time; only last one is effective. For example:

```
Result <= Result xor A(7);
```

- **Exercise 4**

Go back to *Figure 5-14* and fix the multiplexor, not by changing the signal
assignments, but by adding some WAIT statements.

```
ENTITY mux IS
PORT (x, y, select: IN BIT;
      p: OUT BIT);
END mux;
ARCHITECTURE wrong OF mux IS
SIGNAL muxval: INTEGER;
BEGIN
PROCESS
BEGIN
   muxval <= 0;
   IF (select = '1') THEN
       muxval <= muxval + 1;
   END IF;

   CASE muxval  IS
       WHEN 0 => p <= x AFTER 10 NS;
       WHEN 1 => p <= y AFTER 10 NS;
       END CASE;
       wait on x, y, select;
END PROCESS;
END WRONG;
```

-   **Answer to Exercise 4**

Go back to *Figure 5-14* and fix the multiplexor, not by changing the signal
assignments this time, but by adding some WAIT statements.

```
ENTITY mux IS
PORT (x, y, select: IN BIT;
      p: OUT BIT);
END mux;
ARCHITECTURE wrong OF mux IS
SIGNAL muxval: INTEGER;
BEGIN
PROCESS
BEGIN
   muxval <= 0;
wait for 0 ns;  -- force update of muxval
   IF (select = '1') THEN
       muxval <= muxval + 1;
wait for 0 ns;  -- force update of muxval

   END IF;

   CASE muxval  IS
       WHEN 0 => p <= x AFTER 10 NS;
       WHEN 1 => p <= y AFTER 10 NS;
       END CASE;
       wait on x, y, select;
END PROCESS;
END WRONG;
```

The example needs 2 WAIT statements. The first location is after `muxval` gets `0`.
That initializes `muxval`. The second location is after `muxval` gets `muxval + 1`. By
the time you get to the CASE statement on the right-hand-side, the `muxval` has
attained the value from the signal assignment statements, and so, you can proceed
using `muxval`'s current value.   The WAIT statements force the assignment
statements to value `muxval`.

Many designers consider this coding very poor style, and would prefer to avoid this
kind of solution.

- **Exercise 5**

Write the entity declaration in *Figure 5-6* for REG, a 1-bit register.

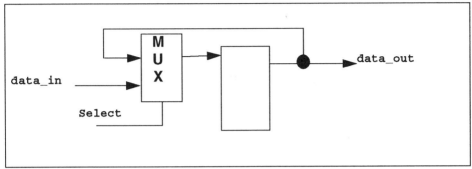

Figure 5-6

- **Answer to Exercise 5**

Write the entity declaration for *Figure 5*-6.

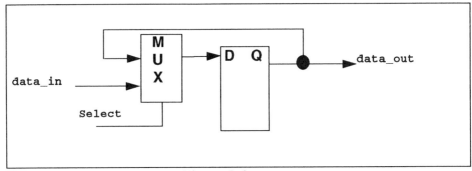

Figure 5-6

```
entity REG is
port (data_in, select: in bit;
      data_out: buffer: bit);
end REG;
```

- **Exercise 6**

Procedure Call - 16-Bit Odd Parity Checker

```
entity RECEIVER is
   port (Y: in BIT_VECTOR (15 downto 0); ODD: out BIT);
end RECEIVER
architecture BEHAVIOR of RECEIVER is
   signal TOP, BOTTOM, ODD: BIT;
begin
   process
   begin
   PARITY (Y(15 downto 8), TOP);
   PARITY (Y(7 downto 0), BOTTOM);
   ODD <= TOP xor BOTTOM;
end BEHAVIOR;
```

Do you see the odd bug lurking here?

-   **Answer to Exercise 6**

Procedure Call - 16-Bit Odd Parity Checker

```
entity RECEIVER is
    port (Y: in BIT_VECTOR (15 downto 0); ODD: out BIT);
end RECEIVER
architecture BEHAVIOR of RECEIVER is
    signal TOP, BOTTOM, ODD: BIT;
begin
    process
    begin
    PARITY (Y(15 downto 8), TOP);
    PARITY (Y(7 downto 0), BOTTOM);
    ODD <= TOP xor BOTTOM;
end BEHAVIOR
```

Do you see the odd bug lurking here?

This is a "killer" bug. Note that the declaration of signal ODD in the architecture makes the entity port signal ODD "invisible". That is, you cannot assign any value to output port ODD!

- **Lab #5**

Moving average exercise

Write a 4-stage moving average process called **P** and place it in an architecture called **moving4**, using type **real** for input port **NEWS** and output port **AVG**. In your process, make sure to declare a new type for the array of real numbers (**real_vect**). This type can be limited to subscripts 1 to 4 or can be unconstrained. Declare a variable array named **A** for the 4 real values being remembered. Use a WAIT statement to get a new result, each **1 ns**.

Analyze your moving average design

1. You may need to declare the range of the input **NEWS**, for example **range 0 to 100**.

2. You may need to declare an initial value on **NEWS** in your entity.

 Now simulate your moving average design, **vhdlsim filter**. If you get a runtime error for initialization, then go back and edit your source above. If your simulation model loads without error, then try the following:

set value on entity input	**NEWS = 53.0**
run 1 ns simulation step	
list values of internals	**avg, NEWS, A**

3. Now try a few more values: **40 40 40 40**. Enter each one, one at a time, and display the results. If it is working OK, continue, else go back and fix it.

4. Edit a script file called **script2** with the commands used above for values:

> **53 40 40 40 40 40 35 35 35 35.**

Execute your script in the simulator

5. Edit your source design replacing the **WAIT** for **1 ns** with a **WAIT ON** signal **NEWS** to change. Rerun and describe how the result differs.

6 | Concurrent Statements

This chapter discusses VHDL concurrent statements. It contains the following sections:

- The Process
- Concurrent Signal Assignment
- Conditional Signal Assignment
- Selected Signal Assignment
- Concurrent Procedure Call
- BLOCK Statement

A hardware description language documents systems that perform parallel operations. During simulation, components in these systems are "run" at the same simulated time, as explained earlier in Chapter 5. Concurrent statements are used to express this kind of parallel behavior. Concurrent descriptions can be *structural* (using components - see Chapter 7) or *behavioral*. The key concurrent statement is the process that is described in detail in Chapter 3.

6.1 The Process

A Process:

- Runs concurrently with other processes.

- Contains only sequential statements.

- Defines regions in architectures where statements execute sequentially.

- Must contain either an explicit sensitivity list or a WAIT statement.

- Provides "programming-language" like capability.

- Can access signals defined in architecture and entity.

Two concurrent processes are shown in *Figure 3-2,* and in the diagram below. The diagram illustrate two processes: Process **A** and Process **B** are running concurrently (i.e., at the same simulated time). Each of them contains a WAIT. The WAIT statement suspends execution of a process. Generally, a simulator works on one process until a WAIT is executed, then on another process.

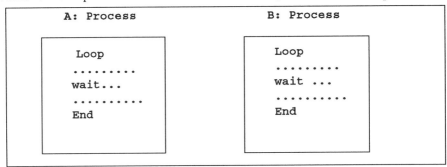

Figure 6-1

The dataflow style of architecture (Section 1.4) utilizes the concurrent statements, which are actually a shorthand way of writing a process. There are several related concurrent statements equivalent to the process statement described in this chapter. Concurrent statements can be organized into blocks to provide hierarchy or to improve readability.

6.2 Concurrent Signal Assignments

Another form of a signal assignment (discussed earlier in Chapter 5) is a concurrent signal assignment, which is used *outside* of a process, but within an architecture. A concurrent signal assignment is a shorthand way to write a process and is equivalent to a process containing one statement. In *Figure 6-2*, the coding on the left contains a concurrent signal assignment to OUTPUT, which is equivalent to the process containing the same statement shown on the right..

Concurrent Form		Sequential Form

`architecture V_1 of V_VAR is`
`begin`
`   OUTPUT <= A (INDEX);`
`end V_1;`
	=	`architecture V_1 of V_VAR is`
		`begin`
		`   process (A, INDEX)`
		`   begin`
		`      OUTPUT <= A (INDEX);`
		`   end process;`
		`   end V_1;`

Figure 6-2

Note, in the concurrent form, that any signal used on the right side of the assignment is by default in the process sensitivity list. Any time a signal changes, the concurrent signal assignment is evaluated and a signal value is assigned.

Architectures can contain several concurrent signal assignment statements, which execute at the same simulation time; there is no order associated with them. *Figure 6-3* shows 2 adders running at the same time; B + C generates A, and E + F generates D. The concurrent signal assignment statements (1 and 2 below) document this behavior. They are equivalent to two processes during simulation. If any of the inputs in the statements change, the statement is executed and a new output value is assigned. Just as in the hardware, the two adders are always running and responsive to their inputs.

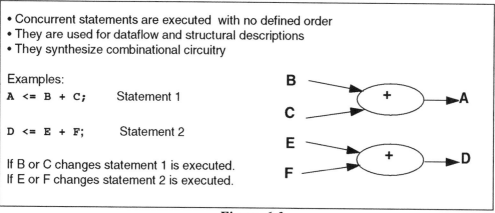

• Concurrent statements are executed with no defined order
• They are used for dataflow and structural descriptions
• They synthesize combinational circuitry

Examples:

`A <= B + C;`   Statement 1

`D <= E + F;`   Statement 2

If B or C changes statement 1 is executed.
If E or F changes statement 2 is executed.

Figure 6-3

Concurrent signal assignments execute asynchronously with no relative order; they are used for data flow descriptions. There are a number of forms for a signal assignment statement, including the AFTER clause and conditional and selected signal assignment.

Forms of Concurrent Signal Assignment

- • Target <= Expression -- logical, comparative, or arithmetic operation

- • Conditional -- similar to IF statements

- • Selected - - similar to CASE statement

6.3 Conditional Signal Assignments

A conditional signal assignment is a concurrent statement and has one target, but can have more than one expression. Only one of the expressions can be used at a time. The syntax is:

conditional_signal_assignment

target **<=** {expression **WHEN** condition **ELSE**} expression;

Example:

```
Z <= A when (x>3) else
     B when (x<3) else
     C;
```

Figure 6-4

In the example, **z** gets **A** or **B** or **C** based on a condition described in a WHEN clause (it is like an IF statement) done outside the process.

Note: You cannot use conditional signal assignment in a process.

Question: What happens when **x** = 3?
Answer:

6.4 Selected Signal Assignments

A selected signal assignment statement can have only one target and can have only one WITH expression. This value is tested for a match in a manner similar to the CASE statement. Below is an example of a selected signal assignment. With **MYSIG** select **A, B,** or **C** based on the current value **MYSIG** (**15, 22**). The syntax is:

selected_signal_assignment

with expression **select**
    target **<=** {expression **when** choices,};

Example:

```
with MYSIG select
    Z <= A when 15,          You can have only one target
         B when 22,
         C when others;
```

Figure 6-5

The above example is like a CASE statement choosing N specific cases. The selected signal assignment is outside of the process. Since it is a concurrent signal assignment, it runs whenever any change occurs to the selected signal in the example **MYSIG**. The selective concurrent signal assignment is a shortcut to writing an equivalent process with a CASE statement.

Note: You cannot use conditional signal assignment in a process.

6.5 Concurrent Procedure Call

Another form of concurrent statement is a concurrent procedure call. A concurrent procedure call is a procedure call that is executed outside of a process; it stands alone in an architecture.

Concurrent Procedure Call:

- Has IN, OUT, and INOUT parameters.
- May have more than one return value.
- Is considered a statement.
- Is equivalent to a process containing a single procedure call.

Figure 6-6 illllustrates a concurrent procedure call. It calls `vector_to_int` with `bitstuff` and `number` in an architecture but not in a process. The example is a concurrent procedure call, exactly equivalent to the process that contains a procedure call statement sensitive to any inputs changing. A concurrent procedure statement is a statement that can run concurrently with other concurrent statements. It is a shorthand way of writing a process around a procedure call.

```
Architecture
begin
vector_to_int (bitstuff, flag, number);      Concurrent Procedure Call
end
```

Equivalent to

```
Architecture
begin
process                                       Process
   begin
   vector_to_int (bitstuff, flag, number);
   wait on bitstuff, number;
end process;
end;
```

Figure 6-6

6.6 BLOCK Statements

A BLOCK statement contains a set of concurrent statements and is especially useful for organizing a design. You can use BLOCK to partition a structural netlist. The syntax is:

block_statement

```
{label:} block [(boolean expression)]
        {declarations}
begin
        concurrent_statements
end block [label];
```

Declarations declare objects local to the block and can be any of the following:

- USE clause
- Subprogram declaration and body
- Type, constants, signals declarations
- Component declaration

Order of the concurrent statements in a block is not significant, since all statements are always executing. Blocks can be nested in your source design file to organize and create hierarchy. Blocks provide name scope. Objects declared in a BLOCK are visible to that block and all blocks nested within. When a *child* block inside a *parent* block declares an object with the same name as one in the parent block, the child's declaration overrides that of the parent. Figure 6-7 shows nested blocks.

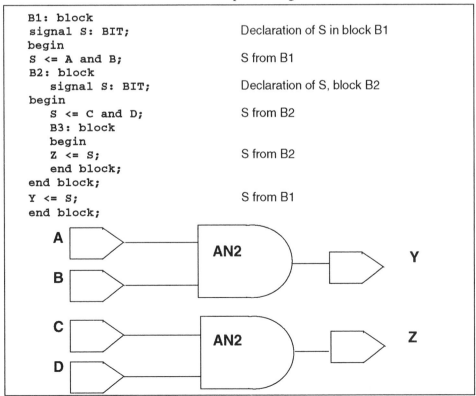

```
B1: block
signal S: BIT;                    Declaration of S in block B1
begin
S <= A and B;                     S from B1
B2: block
    signal S: BIT;                Declaration of S, block B2
begin
    S <= C and D;                 S from B2
    B3: block
    begin
    Z <= S;                       S from B2
    end block;
end block;
Y <= S;                           S from B1
end block;
```

Figure 6.7

Guarded Blocks

A block can have a boolean *guard* expression. This expression automatically creates a boolen signal named **guard**, which you can use to control behavior during simulation. Additionally, a signal assignment statement can have a *guarded clause*. The **guarded** clause makes the execution of the signal asignment statement conditional; assignment is only executed when the signal **guard** is true. This type of signal assignment statement is called a guarded signal assignment statement. An example of guarded signal assignments is presented in Figure 6-8.

```
B1 : block (pdq)              -- Guard expression pdq is true or false
begin
  s <= guarded '1';          -- Executes when guard expression is true
end B1;
```

Figure 6-8

Note: Some synthesis tool vendors do not support guarded blocks. A guarded block is equivalent to a process with a sensitivity list and conditional statements inside the process body. You can use this type of process in lieu of a guarded block.

- **Summary**

1. *Concurrent statements* are statements that execute asynchronously with respect to one another, and at the same simulated time. (Sections 6.2, 6.3, etc.)

2. There are three forms of concurrent statements: the *concurrent assignment statement*, the *concurrent procedure call*, and the *process* itself. Both the concurrent signal assignment statement and the concurrent procedure call statement are each equivalent to a process (Sections 6.2, 6.3, etc.)

3. *PROCESS statement* is a concurrent statement that can contain sequential statements. Inside the process, sequential statements have a program-like meaning. (Section 6.0)

4. Blocks are used to create hierarchy in concurrent sections of an architecture. A block can contain a guard expression.

- **Questions and Answers**

Question: What happens when $x = 3$?

Answer: z gets c

- **Exercise 1**

Conditional Signal Assignment

Syntax:

target <= {expression WHEN condition ELSE} expression;

Exercise:

Write a conditional signal assignment statement to convert:

BOOLY (boolean) to BITTY (bit)   (true becomes '1')

Bitty <=

- **Answer to Exercise 1**

Conditional Signal Assignment

Syntax:

target <= {expression **WHEN** condition **ELSE**} expression;

Exercise:
Write a conditional signal assignment statement to convert:

```
BOOLY (boolean) to BITTY (bit)    (true becomes '1')
```

```
Bitty <=    '1' when BOOLY else '0';
```

- **Exercise 2**

Selected Signal Assignment

Syntax:

```
with expression select
    target <= {expression WHEN choices,};
```

Exercise:

Write the 7 segment decoder using selected signal assignment to LED and from the variable digit.

- Answer to Exercise 2

Selected Signal Assignment

Syntax:

```
with expression select
    target <= {expression WHEN choices,};
```

Exercise:

Write the 7 segment decoder using selected signal assignment to LED and from the variable digit.

```
With digit select
LED <= X"7E" when 0,
    X"6D" when 1,
    X"5C" when 2,
    X"73" when 3,
    X"65" when 4,
    X"37" when 5,
    X"3F" when 6,
    X"62" when 7,
    X"7F" when 8,
    X"77" when 9,
    X"00" when others;
```

-   **Exercise 3**

Comparator Exercise

Build four 1-bit comparators using logical type **bit_vector** for **A**, **B**, and **EQL**, and complete the entity.

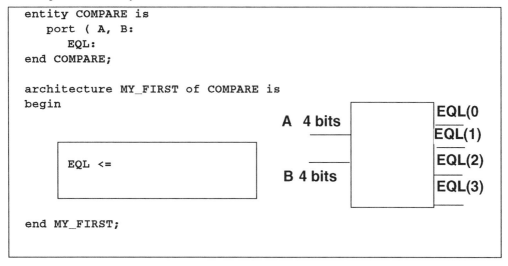

```
entity COMPARE is
   port ( A, B:
      EQL:
end COMPARE;

architecture MY_FIRST of COMPARE is
begin

      EQL <=

end MY_FIRST;
```

- **Answer to Exercise 3**

Comparator Exercise

Build four 1-bit comparators using logical type **bit_vector** for A, B, and EQL, and complete the entity.

```
entity COMPARE is
   port ( A, B: in BIT_VECTOR(3 downto 0);
      EQL: out BIT_VECTOR(3 downto 0));
end COMPARE;

architecture MY_FIRST of COMPARE is
begin
```

```
EQL <= NOT (A XOR B);
```

A 4 bits

B 4 bits

EQL(0
EQL(1)
EQL(2)
EQL(3)

```
end MY_FIRST;
```

- **Exercise 4**

What is Missing?

Below is a one 4-bit Comparator, using logical operators and type BIT_VECTOR.

```
entity COMPARE is
   port (A, B: In BIT_VECTOR (3 downto 0);
         C: out BIT);
end COMPARE;

architecture WOWIE of COMPARE is


begin
   EQL <= NOT (A XOR B);
   C <= EQL (0) and EQL (1) AND EQL (2) AND EQL (3);
END WOWIE
```

```
         4 bits   ┌─────────┐
      A ──────────┤         │
                  │         │
         4 bits   │         ├────────── C
      B ──────────┤         │
                  │         │
                  └─────────┘
```

Hint: Local signals are declared in the architecture declarative part, before the begin.

What is Missing?

Below is a one 4-bit Comparator, using logical operators and type BIT_VECTOR.

```
entity COMPARE is
    port (A, B: In BIT_VECTOR (3 downto 0);
          C: out BIT);
end COMPARE;

architecture WOWIE of COMPARE is

    signal EQL: BIT_VECTOR (3 downto 0);

begin
    EQL <= NOT (A XOR B);
    C <= EQL (0) and EQL (1) AND EQL (2) AND EQL (3);
END WOWIE;
```

Since A, B and EQL are all 4 bits, the signal assignment to EQL produces a 4-bit result, e.g., EQL(0), EQL(1), EQL(2), and EQL(3). EQL needs to be declared as a signal.

- **Exercise 5**

4-Bit Buffer with Enable

Build a set of gates using logical operators.

```
entity GATE is
port (A: In BIT_VECTOR(3 downto 0);
   G: In BIT; C out BIT_VECTOR(0 to 3));
end GATE;

architecture BLIP of GATE is
begin




    ┌──────────────────────────┐              4 bits    ┌─────────────────┐
    │     C <= A and (G);      │          A ─────────   │   GATE          │
    └──────────────────────────┘              1 bit     │                 │ ___C
                                          G ─────────    │   enable        │
    end BLIP                                             │                 │
                                                         └─────────────────┘
```

Draw a gate-level schematic.

What is the problem with the example above?

- Answer to Exercise 5

4-Bit Buffer with Enable

Build a set of gates using logical operators.

```
entity GATE is
port (A: In BIT_VECTOR(3 downto 0);
   G: In BIT; C out BIT_VECTOR(0 to 3));
end GATE;

architecture BLIP of GATE is
begin
```

```
C <= A and (G&G&G&G);
```

```
end BLIP;
```

Draw a gate-level schematic.

What is the problem with the example above?

G needs to expand to 4 bits -- for example, using concatenation.

- **Exercise 6**

Relational Exercise

Build a 4-bit comparator using relational operators.

```
entity COMPARE is
port (A, B: In BIT_VECTOR(3 downto 0); C out BOOLEAN);
end COMPARE;

architecture BETTER of COMPARE is
begin



C <=

  end BETTER
```

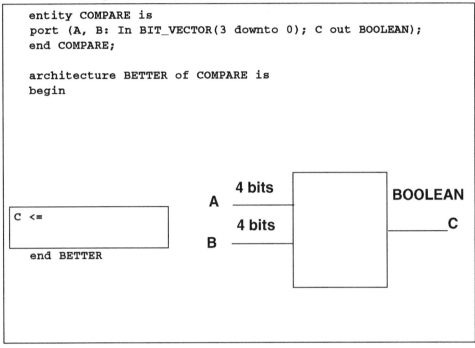

-   **Answer to Exercise 6**

Relational Exercise

Build a 4-bit comparator using relational operators.

```
entity COMPARE is
port (A, B: In BIT_VECTOR(3 downto 0); C out BOOLEAN);
end COMPARE;

architecture BETTER of COMPARE is
begin
```

```
C <= (A = B);
```
```
end BETTER;
```

Concurrent signal assignment is continuously executing in the hardware.

- **Exercise 7**

BCD Decoder Diagram

Decode D(3) downto D(0) producing 4 signals: NINE, EIGHT, ONE, ZERO.

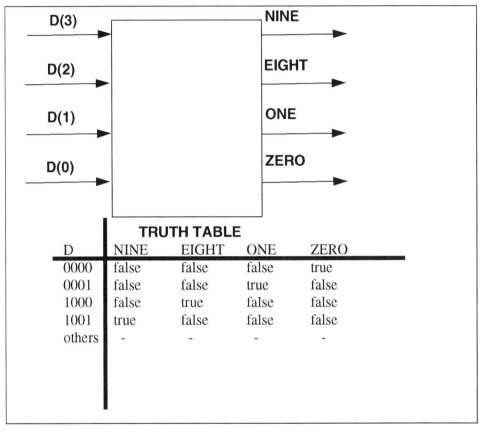

	TRUTH TABLE			
D	NINE	EIGHT	ONE	ZERO
0000	false	false	false	true
0001	false	false	true	false
1000	false	true	false	false
1001	true	false	false	false
others	-	-	-	-

```
Entity BCD_DECODER is
port (D: _____;
   NINE, EIGHT, ONE, ZERO: out BOOLEAN);
end BCD_DECODER

architecture MY_FIRST of BCD_DECODER is

begin
         _____;
         _____;
         _____;
         _____;
```

- **Answer to Exercise 7**

BCD Decoder Diagram

Decode D(3) downto D(0) producing 4 signals: NINE, EIGHT, ONE, ZERO.

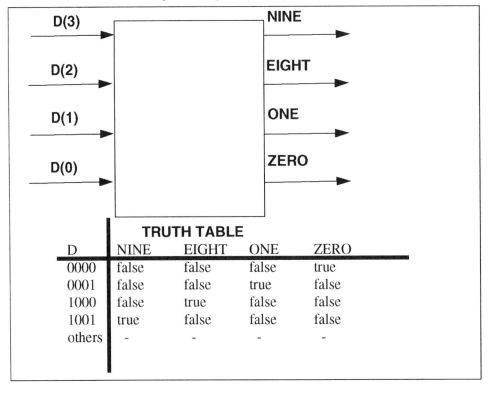

TRUTH TABLE

D	NINE	EIGHT	ONE	ZERO
0000	false	false	false	true
0001	false	false	true	false
1000	false	true	false	false
1001	true	false	false	false
others	-	-	-	-

```
Entity BCD_DECODER is
port (D: in BIT_VECTOR (3 downto 0);
   NINE, EIGHT, ONE, ZERO: out BOOLEAN);
end BCD_DECODER;

architecture MY_FIRST of BCD_DECODER is

begin
   NINE <= (D = "1001");
   EIGHT <= (D = "1000");
   ONE <= (D = "0001");
   ZERO <= (D = "0000");
-- no other cases decoded
end BCD_DECODER;
```

- **Lab #6**

Moving average using signals

Write a 4-stage moving average architecture without a process and place it in an architecture called `moving4`, using type `real` for input port `NEWS` and output port `AVG`. Make sure to declare a new type for the array of real numbers (`real_vect`). This type can be limited to subscripts 1 to 4, or it can be unconstrained. Declare a signal array named `A` for the 4 real values being remembered.

Now analyze your moving average design:

1. You may need to declare the range of the input `NEWS` (for example, range 0 to 100).

2. You may need to declare an initial value on `NEWS` in your entity.

Now simulate your moving average design, `vhdlsim filter`. If you get a runtime error for initialization, then go back and edit your source above. If your simulation model loads without error, then try the following:

- Set the value on entity input `NEWS` to 53
- Run 1 ns simulation step
- List values of internals

3. Now try a few more values: 40, 40, 40 , 40. Enter one at a time and display the results. If your design is working, continue. Else, go back and fix it.

4. Edit a script file called `script2` with the commands used above for values: 53, 40, 40, 40, 40 , 40, 35, 35, 35, 35.
Execute your script in the simulator type: include `script2`.

5. Edit your source design replacing the `WAIT for 1 ns` with a `WAIT ON` signal `NEWS` to change. Rerun and describe how the results differ.

7 | Structural VHDL

This chapter discusses the structural style of VHDL. It contains the following sections:

- Component Instantiation
- Generate Statement
- Configurations
- Generics

The VHDL structural style describes the interconnection of components within an architecture. In a structural architecture, you declare the components that you are using, then create instances of those components with particular mappings of signal wires to the various pins of the components. Component instantiation statements identify wired connections. Each component instantiation is a concurrent statement similar to those described in Chapter 6. Configurations allow you to select particular versions of a component (e.g., different speeds or power levels).

Structural style is similar to a netlisting language in other CAD systems. Hence, the order of the components is not important. The syntax is:

component_declaration

   component component_name **port** (list) **end component;**

component_instantiation

   label: component_name **port map** ([named | positional]);

Most top-level designs utilize the structural style where lower-level design elements are previously compiled as components in a library. See Sections 1.4 and 5.1.

Note: Entities and architectures cannot be nested in a VHDL source design.

The figure below describes a flip-flop using two-input NOR gates.

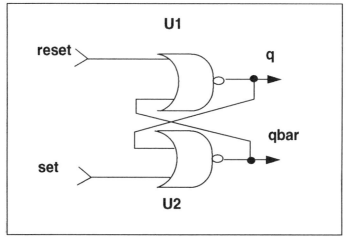

Figure 7-1

There are two components designated in the schematic in *Figure 7-1*: component
U1 and component **U2**. The input signals coming into the flip-flop entity are **set**
and **reset** signals; **q** and **qbar** are the output signals. These signals connect to the
nor2 components, whose pins are designated with inputs **a** and **b** and output **c**, as
shown in *Figure 7-2*.

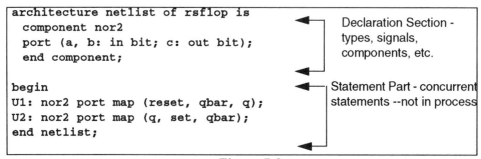

Figure 7-2

The statement part of the architecture designates the signals connected for the **U1**
and **U2 NOR2** gates. Note that components are used in an architecture and not in a
process.

As an exercise, complete the entity description for the **rs_flop** shown in the
schematic in *Figure 7-1*.

```
entity rs_flop is
   port (                                                           );
end rs_flop;
```

q and qbar are of type inout in the entity description *Figure 7-3* because the q is an output of the flip-flop and is also used inside a component that is referring to a signal.

A process is a behavioral description and cannot contain components. Component instances use signals that are previously declared:

- As entity ports
- Signals declared in an entity
- Signals declared in a package
- Signals declared in an architecture (see Chapter 5)

You can also have function calls in place of pin names. This substitution is useful for doing type conversion when required in component hookups. If no connection is to be specified, use the reserved VHDL word OPEN.

The solution is:

```
entity rs_flop is
    port (set,reset: in bit; q, qbar: inout bit);
end rs_flop;
```

Question: Why are q and qbar inout?
Answer:

7.1 Component Instantiation Using Named Notation

In addition to the positional association of signals illustrated in the port map of the component instantiation statement, VHDL supports named association, as shown in the next example.

```
U1: nor2                                    Positional Association
port map (reset, qbar, q);                  same as


U1: nor2                                    Named Association
port map (a => reset, c => q, b => qbar);
```

Figure 7-3

In this case, you associate the actual signal being connected to a pin, with each of the original names of signal pins used in the component declaration. A right arrow => reads as *gets*, i.e., pin a gets signal reset. The use of named association makes VHDL more readable and helps you remember which pin is associated with which signal. You can specify pins in any order (with positional order you cannot).

7.2 Generate Statement

To facilitate describing arrays of components, you can use the GENERATE statement. A GENERATE statement is a type of loop that can be used outside of a process. It is like a LOOP statement, which allows an iteration of components. The GENERATE statement provides the ability to replicate structures outside a process. It also allows for special conditions for IF and THEN statements. It is useful in applications such as memory arrays and register arrays. The syntax is:

generate_statement

    label: {[FOR_specification

         IF_condition]} **generate**

      {concurrent_statements}

    end generate;

Before looking at the GENERATE statement, let us explore specifying a 4-bit shift register the long way:

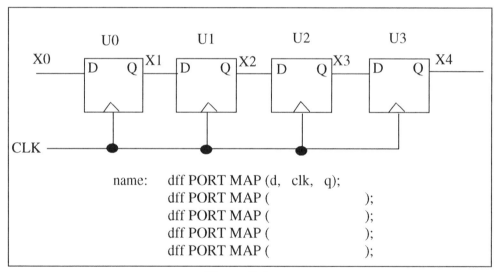

Figure 7-4

Question: What values should be put in for **PORT MAP** above?

The GENERATE statement is convenient for expressing an array of components with similar connections. It reduces the amount of typing you need to do for original design entry.

In the next example, the 4 D-flops are connected with a loop `for i IN 0 TO 3`. Following this code is an expression involving indexes, `X(i)` and `X(i+1)`.

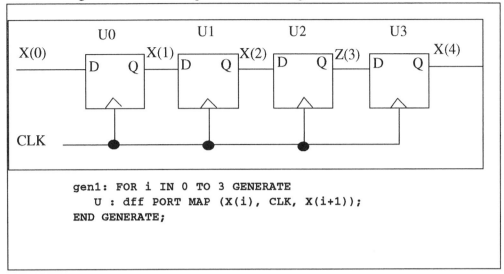

```
          gen1: FOR i IN 0 TO 3 GENERATE
            U : dff PORT MAP (X(i), CLK, X(i+1));
          END GENERATE;
```

Figure 7-5

This code generates 4 statements equivalent to *Figure 7-5*.

Figure 7-6 illustrates an entity `shift` and architecture `gen_shift`   that generate a shift register. It includes a GENERATE statement with a flip-flop.

```
ENTITY shift IS
   PORT (SIN, CLK: IN bit; SOUT: OUT bit);
END shift;

ARCHITECTURE gen_shift OF shift IS
   COMPONENT dff
      PORT (d, clk: IN bit;q: OUT bit);
   END COMPONENT;
   SIGNAL X: bit_vector (0 to 4);
BEGIN
   X(0) <= SIN;
   gen1: FOR i IN 0 TO 3 GENERATE
     U: dff PORT MAP (X(i), CLK, X(i+1));
   END GENERATE;
   SOUT <= X(4);
END gen_shift;
```

Figure 7-6

You do not always want to generate every slice of design in an identical way. In *Figure 7-7* notice that the first flip-flop and last flip-flop are distinguished in that they do not interconnect to other flip-flops on their ends. Therefore, you need a special condition for the first and last element of this array.

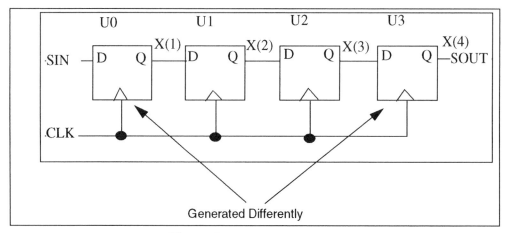

Figure 7-7

The first part of *Figure 7-8* illustrates the use of the GENERATE statement with some IF statements. The first case is for i = 0. The middle case is for i = 1 and 2 and the last case is for i = 3 . This example illustrates the use of the GENERATE statement with some special testing conditions. It separates the generation of those two pieces designated to the left side and the right side.

```
FOR i IN 0 TO 3 GENERATE
IF (i = 0) GENERATE
   UA : dff PORT MAP (SIN, CLK, X(i + 1)); END GENERATE;   Left Side

IF ((i > 0) AND (i < 3)) GENERATE
   UB : dff PORT MAP (X(i), CLK, X(i + 1)); END GENERATE; Inside

IF (i = 3) GENERATE
   UC : dff PORT MAP (X(i), CLK, SOUT); END GENERATE;      Right Side

END GENERATE;
```

Figure 7-8

7.3 Hierarchy

A VHDL design consists of a hierarchy of connected components compiled from behavioral, dataflow, or structural architectures. *Figure 7-9* shows a design hierarchy of COMPARE using connected components. Once designs are compiled, they go into the library and can be used as components in subsequent designs. You may declare a component before you take an instance of it from a library. A design containing components requires a configuration (see Section 7.4).

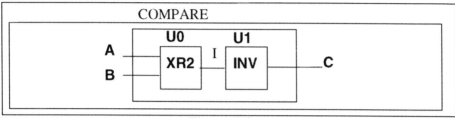

Figure 7-9

Component Declarations

Figure 7-10 highlights two architectures of COMPARE. They are in a structural style of architecture. The difference between them is that in s, on the left, components XR2 and INV are declared before they are used (required in VHDL87, optional in VHDL92). In architecture T on the right, these component declarations occur in a package called xyz_gates residing in library WORK. Architecture T has a USE statement to access the component declarations; it is a shorter design. This example illustrates using component declarations in a package. You can use packages from a library to hold frequently-used declarations.

```
architecture S of COMPARE is        use work.xyz_gates.all;
signal I: bit;                       architecture T of COMPARE is
component XR2 port (X,Y: in bit;     signal I: bit;
                Z: out bit);         begin
end component;                          U0: XR2 port map (A, B, I);
component INV port (X: in bit;          U1: INV port map (I, C);
                Z: out bit);         end T;
end component;
begin
   U0: XR2 port map (A, B, I);
   U1: INV port map (I, C);
end S;
```

Figure 7-10

Figure 7-11 shows the VHDL source design unit of the `package xyz_gates`. There are two component declarations in the package: `XR2` and `INV`. Notice that the package saves you from having component declarations elsewhere. The components themselves must have an entity and an architecture and must be compiled in a library. The package only contains the component declaration (see Chapter 8 on packages).

```
package xyz_gates is


component XR2
port (X, Y: in bit; Z: out bit);
end component;


component INV
port (X: in bit;  Z: out bit);
end component;


end xyz_gates;



-----no package body-----------
```

Figure 7-11

Figure *7-12* is a description of design entity `XR2`, which can be used as a component after it has been compiled into a library. There are two architectures of `XR2` in this case. There is an architecture called `slow`, with a delay time of `1 ns` and an architecture `fast` with a delay of `.4 ns`. This example illustrates that for an entity you can have more than one architecture compiled, and the architectures must have different names. In this example, they have different speeds. When you use the `XR2` component entity in a design, you get, by default, the most recently compiled (latest) architecture of that entity in the library (e.g., `fast`).

```
entity XR2 is
   port (X, Y: in bit;  Z: out bit);
end XR2;

architecture slow of XR2 is
   begin
      Z <= X xor Y after 1.0 ns;
end slow;

architecture fast of XR2 is
   begin
      Z <= X xor Y after 0.4 ns;
end fast;
```

Figure 7-12

7.4 Configurations

Design entities can have more than one architecture, as illustrated in the two architectures of **XR2** in *Figure 7-12* (also see **COMPARE** in Chapter 1). A CONFIGURATION statement is used to select a particular architecture of an entity. A *configuration* is a convenient way of documenting different versions of a design during the design process. It also saves you time because you do not need to recompile the entire design if you need to substitute only a few components. A selection of architectures can be coded <u>within</u> an architecture or can be a separately compiled configuration design unit.

Configuration in Architecture

When you have architectures that use other design entities as components, you can select a particular architecture (i.e., a CONFIGURATION) as shown in *Figure 7-13* as part of the architecture of entity **COMPARE**.

```
architecture STRUCTURAL of COMPARE is
......
for U0: XR2 use entity work.XR2(fast);                    Configured
begin
U0: XR2 port map (A, B, I);   ◄
U1: INV port map (I, C);◄───────────── Not configured
end STRUCTURAL;
```

Figure 7-13

The example is configuring the **XR2** gate named **U0** in the architecture **STRUCTURAL**. This configuration specification of the **XR2** gate has a FOR/USE statement. Notice:

```
for U0: XR2 use entity work.XR2(fast);
```

The entity **XR2** and the architecture **fast** are selected from library **work**. The inverter **U1:INV** has not yet been configured in this design. A configuration particularizes a design. As previously mentioned, a configuration can be coded inside of an architecture (as shown in this example) or can stand alone and be a design unit. Obviously, there are advantages of having a configuration clause in an architecture and there are advantages of compiling configurations separately.

Configuration Outside of an Architecture

A configuration is a design unit that can be compiled separately and stored in a library. It creates a simulation object. It also allows you to particularize and create a variation instance of a design. *Figure 7-14* defines a design unit **wonderful**, which is a version of **COMPARE** that uses the **fast XR2** component. **wonderful** can now be simulated (see *Figure 7-12* for the declaration of the architecture **fast**).

```
configuration wonderful of COMPARE is
   for structural
      for U0: XR2 use entity work.XR2(fast); end for;
   end for;
end wonderful;
```

Figure 7-14

Figure 7-15 is an example, which shows the explicit configuration of the **XR2** gate as a separately compiled design unit. It configures a version of the XR2 named configuration **speedy**. A configuration statement simply names the architecture to be used, which is sufficient for specifying a named configuration. Using **CONFIGURATION Speedy** is the same as selecting architecture **fast** of the entity **XR2** shown in *Figure 7-12*.

```
CONFIGURATION Speedy OF XR2 IS
    FOR fast                    -- Architecture to be used
    END FOR;
END Speedy;
```

Figure 7-15

You might want to give your configuration a separate name when you wish to distinguish a particular version of the entity by using a cataloged name.

Two Ways of Configuring

Notice in *Figure 7-16* that there really are two ways of configuring the **XR2** in the configuration **wonderful** in *Figure 7-14*. You can refer to an entity name with an architecture name (e.g., **XR2(fast)**) or you can name a previously compiled configuration (e.g., **speedy** in *Figure 7-15*).

```
for U0: XR2 use entity work.XR2(fast); END FOR;
                                                          same as
for U0: XR2 use CONFIGURATION WORK.speedy; END FOR;
```

Figure 7-16

Configuration Options

It is not necessary to configure every component uniquely by instance name. You can use *all* or the *other* option. The figure shows examples of how you can configure the **XR2** and **INV** components. Note the use of **others** and **all** in *Figure 7-17*. You can either call out the unique instance names first, followed by the others option <u>or</u> use the all option.

```
CONFIGURATION wonderful OF COMPARE IS
for structural
   for U0: XR2 use CONFIGURATION WORK.speedy; end for;
   for others: XR2 use CONFIGURATION WORK.creepy; end for;
   for all: inv use CONFIGURATION WORK.tiny; end for;
   end for;
end wonderful;
```

Figure 7-17

Configuration **wonderful** is a separately compiled design unit of architecture **structural** of **COMPARE**. It must be compiled after configurations **speedy**, **creepy**, and **tiny** (the last two are not shown).

Port Remapping

Another feature of configurations is that you can remap components and ports.

```
CONFIGURATION nandy OF compare IS
   FOR structural
      FOR all: inv USE ENTITY WORK.nand2(behave)
         PORT MAP (a=>a, b=> VCC, z=>z);
      END FOR;
   END FOR;
END nandy;
```

Figure 7-18

Figure 7-18 shows an interesting application of configurations. An inverter has 1 input and 1 output. A nand2 gate has 2 inputs and 1 output. You can, in some cases, substitute a nand2 gate for an inverter. You may need to create this variation of an architecture if you run out of inverters and want to build a variation of the design using NAND gates instead. This example illustrates the configuring of an inverter using a NAND gate. Notice that it says, as part of the configuration, for all inverters **inv** use **WORK.nand2** (the behavioral model). You use a NAND gate for the inverter in all cases. There is, however, an extra pin. The code says to map port **a** of the inverter to port **a** of the NAND gate , and map port **b** of the NAND gate to **vcc**. The example code ties the pin to a constant **vcc**. The output **z** of the NAND gate is connected to the output Z of the inverter.

The example illustrates a substitution of a part and a remapping of the pins. The configuration not only allows you to choose a particular architecture, it also allows you to choose a different kind of component with a similar function, and to do some port mapping at the same time.

Default Configuration

When the top level architecture contains components, a configuration design unit is required in most simulation systems. The simplest configuration needs only to identify the entity and architecture name. For example:

```
configuration cfg_compare of COMPARE is      entity name
   for structural                            architecture name
   end for;
end cfg_compare;
```

The latest version of the architectures are found in library WORK.

Note: When a simulation run is made, remember to simulate the configuration name. For example:

```
vhdlsim cfg_compare
```

Also see Chapter 1.

When a design does not contain an explicit configuration specification for an instantiated component (naming the library and entity name), a previously compiled entity with the same name as the component is located (bound) by default. The library WORK, in addition to any libraries that are visible via a USE clause, are searched to find an entity with the same name as the component.

Null Configuration

When an architecture does not contain components, it does not need an explicit configuration. The most recently compiled architecture of an entity is the configuration used -- this is sometimes referred to as the *null configuration*.
See *Figure 8-1* for a null configuration of COMPARE.

7.5 Generics

In addition to having input/output ports in an entity, an entity can specify parameters (generic declarations) that allow you to alter a design's behavior when you instantiate the component, using a *generic map*. Additionally, a generic parameter can also have a default value.

Generic Declaration

Figure 7-19 illustrates how to declare a component entity with a generic parameter and a default value. The parameter m specifies a delay time in the example.

```
entity XR2 is
   generic (m:time:= 1.0 ns);                    delay time = m
   port (X, Y: in bit; Z: out bit);
end XR2;

architecture GENERAL of XR2 is
   begin
   Z <= X xor Y after m;                         generic used
end general;
```

Figure 7-19

Note that the generic declaration must come before the port declaration in an entity. In *Figure 7-19* the generic is of type time and it has a default value of 1 ns. The architecture GENERAL uses the generic parameter m to establish a delay value for z. The generic parameter m controls the timing behavior of the XR2 model, when the XR2 is simulated.

Generic Map in Component Instance

Figure 7-20 illustrates using the component XR2 and passing a parameter value via a generic map for the delay value m, 1.5 ns in this case.

```
architecture STRUCTURAL of COMPARE is
signal I:bit;
component XR2
   generic (m:time);
   port (x, y: IN BIT; z: OUT BIT);
end component;

begin
U0: XR2 generic map (m => 1.5 ns)              generic map
        port map (I, A, B);
.......
end STRUCTURAL;
```

Figure 7-20

You must specify the `generic map` clause before the `port map`. The order is important; you must <u>not</u> have a semicolon separating the two. You assign a value to the parameter `m` with a right arrow. This code reads as `m` gets `1.5 ns`. The code takes an instance of the `XR2` gate, previously compiled as a component in a library, and provides a parameter, which is the delay time. This delay time overrides the originally specified default of `1 ns`. If you did not specify a generic map, the value of `m` would get the default value. This example illustrates a parameterized design using a single generic parameter. There can be more than one generic parameter and they can be of any previously declared type.

Note: `generic map` before `port map`, with no `;` separator.

Generic Default in a Component Declaration

Figure 7-21 shows the component declaration of `XR2` within a comparator.

```
architecture STRUCTURAL of COMPARE is
signal I: bit;
component XR2
   generic (m: time:= 1.5 ns);          default value
   port (x, y: IN BIT; Z: OUT BIT);
end component;

begin
   U0: XR2 port map (A, B, I);          no generic map
   . . . . . . . . . .
end STRUCTURAL;
```

Figure 7-21

Inside the architecture `COMPARE` is a component declaration with a generic parameter `m`, which specifies the default parameter for all of the `XR2`s in the architecture. Notice that `U0` does not provide a generic map; the `generic` value is provided by the component declaration default value.

Note: You can only have one default value; it can either be in the component declaration or in the entity declaration.

You can override a default value by using a generic map in the instance.

Generic Default in a Package

For convenience, one of the best places for specifying the default values of generic parameters is in a package in which the components are declared. *Figure 7-22* shows a generic default value as part of the component declaration in a package.

```
package comp_unit_delay is        --generic declared in component declaration


  component XR2
    generic (m: time:= 1.0 ns);--delay time
    port (x, Y: IN BIT;  z: OUT BIT);
  end component;


  component INV
  ........
  ........
  end comp_unit_delay;
```

Figure 7-22

The **XR2** component declaration is in the package, including the generic parameter **m** and default value. The package provides a default value for the generic when the component is instantiated. A generic can only have one default value.

A generic is a general mechanism for passing instance-specific data into a component. The generic provides a parameter similar to the way that a VHDL passes a parameter to a function. The data passed in is constant and cannot be modified when it is used. Using generics is a very good way to specify delay values, load values, and other instance-specific parameter values. In addition to specifying generic values in component declarations and in component instances, you can provide generic parameter value information as part of a configuration.

Mapping Generics in a Configuration

Figure 7-23 illustrates that you can configure **COMPARE** and provide a generic map of parameter **m**. The time delay for **XR2** is **1.2 ns**. One of the ways you can configure a design is to parameterize the timing of a design. Thus, every configuration is really just a variation of the timing parameters. You can create many configurations of a design; each configuration gives a different generic map.

```
  CONFIGURATION late OF compare IS
    FOR structural
      FOR U0: XR2 USE ENTITY WORK.XR2 (XR2_gen1)
        GENERIC MAP (m => 1.2 ns);
      END FOR;
      ...........
    END FOR;
  END late;
```

Figure 7-23

Behavioral Model with Generics

You can use generics outside of structural architectures. You can also use generics in behavioral architectures. Recall the typical conversion equation, of the form $y = mx + b$, used in the function c_to_f. *Figure 7-24* illustrates that the value of m, b, the so-called parameters of the function, can also be supplied as generics. In the example is a design entity called linear which takes in the value x and gives us the value y and transforms it according to parameters supplied in the generics. The example shows how you can declare the generics as part of the entity linear.

```
ENTITY Linear is
    generic (m, b: real);
    port (W: in real, Y: out real);
end Linear;

Architecture one of Linear is
Function c_to_f (W:real;) return real is
    begin
    Return (m * W) + b;
end c_to_f;

Begin
    ...
    ...c_to_f(...                              Function Call
    ...
end one;
```

Figure 7-24

- **Summary**

1. VHDL *structural style* describes the interconnection of components within an architecture. It is similar to a netlisting language in other CAD systems. (Section 7.0)

2. In a *structural architecture*, you declare the components, then create instances of those components with particular mappings of signal wires to the various core pins of the components. (Section 7.0)

3. The GENERATE statement provides the ability to replicate structures and concurrent statements outside a process. It also allows for special conditions for IF and THEN statements. It is useful in applications such as memory arrays and register arrays. (Section 7.2)

4. Configurations bind particular entities to component instantiations. They can be used to bind architectures to configurations. (Section 7.4)

5. Configurations connect values to component instances either by binding, by choosing, or by selecting a particular value of a parameter. These values can be generic parameters, different component variations, or different pin configurations. (Section 7.4)

6. Configurations let you give instance-specific information when using components and allows you to save compiled variations under different named configurations in a library. (Section 7.4)

7. Generics allow parameters to vary function in component instance, component declaration, or configuration. (Section 7.5)

- Questions and Answers

Question: Why are q and qbar inout?

Answer: Because they exit the design and are also used within the design.

Question: What values should be put in for port map?

Answer:

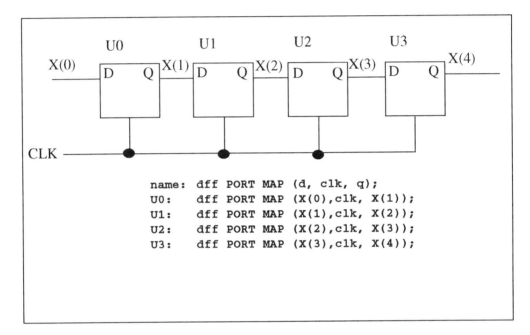

```
name:   dff PORT MAP (d, clk, q);
U0:     dff PORT MAP (X(0),clk, X(1));
U1:     dff PORT MAP (X(1),clk, X(2));
U2:     dff PORT MAP (X(2),clk, X(3));
U3:     dff PORT MAP (X(3),clk, X(4));
```

-   **Exercise 1**

Complete the specification below, using generics for parameters m and B.

```
ENTITY Linear is



      port (W: in real, Y: out real);
end Linear;
Architecture only of Linear is
Function c_to_f (w:real) return real is
   begin
   Return (M * W) + B;
end c_to_f;
   .

   .
end Linear;
```

- **Answer to Exercise 1**

Complete the specification below, using generics for parameters m and B.

```
ENTITY Linear is

   generic (m, B: real);

    port (W: in real, Y: out real);
end Linear;
Architecture only of Linear is
Function c_to_f (W:real) return real is
   begin
   Return (M * W) + B;
end c_to_f;
   .
   .
   .
end Linear;
```

- **Lab #7.1**

A BCD to 7 segment lamp converter needs to be completed. The entity DECODER is described below. Complete its architecture. The entity CONVERTER is completed below. Complete the designs BCD7, which connects the DECODER to the CONVERTER.

```
entity CONVERTER is
   port (NINE, EIGHT, SEVEN, SIX, FIVE, FOUR, THREE, TWO, ONE, ZERO:
   in bit;
      A, B, C, D, E, F, G: out bit);
end CONVERTER;

architecture BEHAVIOR of CONVERTER is
begin
   A <= ZERO or ONE or TWO or THREE or FOUR or SEVEN or EIGHT or NINE;
   B <= ZERO or ONE or THREE or FOUR or FIVE or SIX or SEVEN or EIGHT
   or NINE;
   C <= ZERO or TWO or THREE or FIVE or SIX or EIGHT or NINE;
   D <= ZERO or TWO or SIX or EIGHT;
   E <= ZERO or FOUR or FIVE or SIX or EIGHT or NINE;
   F <= ZERO or TWO or THREE or FIVE or SIX or SEVEN or EIGHT or NINE;
   G <= TWO or THREE or FOUR or FIVE or SIX or EIGHT or NINE;
end BEHAVIOR;
```

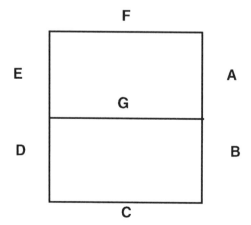

```
entity BCD7 is
port (NIBL: IN BIT_VECTOR (3 downto 0);
            SEGMENTS: out BIT_VECTOR (1 to 7));
end BCD7;

architecture BEHAVIOR of BCD7 is
signal S0, S1, S2, S3, S4, S6, S7, S8, S9: bit;
component DECODER
  port (BCD: in BIT_VECTOR (3 downto 0);
  NINE, EIGHT, SEVEN, SIX, FIVE, FOUR, THREE, TWO, ONE, ZERO:out bit);
end component;

conponent CONVERTER
   NINE, EIGHT, SEVEN, SIX, FIVE, FOUR, THREE, TWO, ONE, ZERO:
end component;

begin
U0: DECODER port map (NIBL, S9, S8, S7, S6, S5, S4, S3, S2, S1, S0);
U1: CONVERTER port map (                                       );
end BEHAVIOR;
```

```
entity DECODER is
port (
  BCD: in BIT_VECTOR (3 downto 0);
  NINE, EIGHT, SEVEN, SIX, FIVE, FOUR, THREE, TWO, ONE, ZERO:
        out BOOLEAN
end DECODER;

architecture BEHAVIOR of DECODER is
begin



end BEHAVIOR;
```

- **Lab 7.2**

Specify a 4-bit loadable register **REG4**. Use the flip-flop component described below:

`FD1(D, CP, Q, QN)`

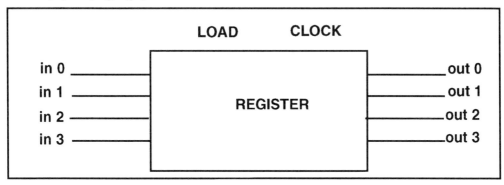

Note: When load is false, the outputs of the flip-flops need to recirculate to the inputs through a multiplexor. Provide this multiplexing with a logic function:

`D <= IN when (LOAD = '1') else OUT;`

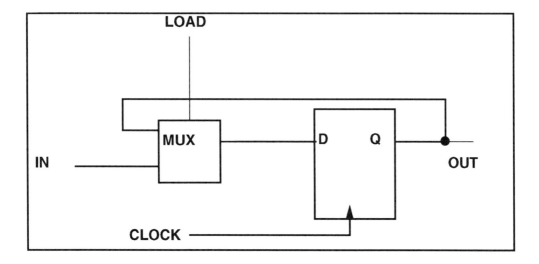

4-Bit Loadable Register

```
entity RED4 is
port (INSTUFF: in BIT_VECTOR(3 downto 0);
      OUTSTUFF:out BIT_VECTOR(3 downto 0);
end REG4;

architecture BEHAVE of REG4 is
   component FD1 port (D, CP: in BIT; Q, QN: out BIT);
   end component;
   signal d, q, qn: BIT_VECTOR (3 downto 0);
   ┌─────────────────────────────────────────┐
   │                                         │
   └─────────────────────────────────────────┘
end BEHAVE;
```

Note: in and out are reserved words and the entity port names need changing.

8 | Packages & Libraries

This chapter discusses packages and libraries. It contains the following sections:

- Deferred constant
- Subprogram declaration
- Component declarations
- Selected name and the USE statement
- General notes on packages
- Typical vendor packages
- IEEE Package 1164

A package is a compilable VHDL source design unit, written by a vendor or a user, which collects frequently used declarations. Once compiled into a library, you can call out a group of declarations with a USE statement or by using selected names to get a particular item. Packages usually contain type declarations, constant declarations, and subprogram declarations. Some examples of these declarations are shown in previous chapters (see Chapters 2 and 4). A package allows you to share common information in different designs and isolate changes.

In particular, a package consists of two compilable parts: *a package header* declaration and a *package body* declaration. A header declaration declares the names of things that are declared in the package body declaration. In some cases, the header can give the actual values and substance to the things being declared. You must compile the package header before you compile the corresponding package body. A package header and a package body contain the following:

- Package Header
    - subprogram declarations
    - type declarations
    - component declarations
    - deferred constant declarations

- Package Body
    - subprogram body
    - deferred constant value

The separation of header and body in separate files allows isolating changes and minimizing the number of other recompiled design units.

8.1 Constant Declarations

Below is a package `my_defs` that contains a constant called `unit_delay`. The package header gives the value of 1 `ns` to `unit_delay`, and there is no package body. After `my_defs` is compiled into `freds.library`, `unit_delay` can be used in any other design. The architecture `flow` is separately compiled and can access `my_defs` to get the value `unit_delay` because of the USE clause (see Section 8.5).

```
Package my_defs is
   Constant unit_delay: time := 1 ns;
end my_defs

entity COMPARE is
   port (a, b : in bit; c : out bit);
end COMPARE;

library freds_library;
Use freds_library.my_defs.all;              USE clause

architecture flow of COMPARE is
begin
   c <= NOT (a XOR b) after unit_delay;
end flow;
```

Figure 8-1

One of the problems with this type of constant declaration is that if you change the constant value (1 ns) you need to recompile the package and any design that refers to it (e.g., architecture `flow`). Typically, a VHDL simulator issues a warning message if the package was compiled after a design unit referred to it.

8.2 Deferred Constants

Instead of recompiling (as in *Figure 8-1*), you can use a *deferred constant as shown* in *Figure 8-2*. The package header declares a constant called `unit_delay`. Its actual value is deferred to the package body. The types must match. When a simulation run occurs, the value of `unit_delay` is obtained from the package body. You can recompile the package body and change its value without recompiling the package header. Any designs referring to the constant `unit_delay` do not need to be recompiled. You can change a constant value by compiling only the package body.

Note: Put the package header and the package body in separate files to make them separately compilable.

```
PACKAGE my_defs IS                          Package Header
   constant unit_delay :time;               Value deferred
END my_defs;

PACKAGE BODY my_defs IS                      Package Body
   CONSTANT unit_delay: time := 1.0 ns;
END my_defs;
```

Figure 8-2

8.3 Subprograms in Packages

The example shows a function that is declared in a package. In this case, you must declare, in the package header, the interface to the functions (i.e., the names of the functions and the arguments). The complete function declaration occurs in the package body. In *Figure 8-3*, there are two functions in the package `tempstuff`: `C_to_F` and `F_to_C`. These functions are temperature conversions that operate on real numbers, as shown in the package header. The package body contains the entire function declaration.

Header and body are separately compiled. There is a compilation order: the package header needs to be compiled before the package body. If you recompile the package header, you must recompile the body, along with any design that uses the package functions. If you change the algorithm in the package body, you do not need to recompile the package header if the arguments are the same. Any architecture using the function does not have to be recompiled. This feature is the advantage of having a separate package header and package body. As long as arguments are the same, only the package body needs to be recompiled.
Note: You should keep header and body in separate source files.

```
PACKAGE tempstuff IS                              Package Header
Function C_to_F (C: real) return real:
Function F_to_C (f: real) return real;
end tempstuff;


Package Body tempstuff IS                         Package Body

Function C-to_F (c: real) return real is
begin
Return ( c * 9.0/5.0) + 32.0;
end C-to_F;

Function F_to_C (f: real) return real is
..........
end tempstuff;
```

Figure 8-3

8.4 Component Declarations

As mentioned previously, rather than having the component declarations in line in a structural design, it is more convenient to have the component declarations in a package (see *Figure 7-10*). *Figure 8-4* illustrates component instantiations in an architecture using declarations in a package. In the example below there is a library `xyz`, which contains a compiled package named `components`. The USE statement identifies the library name `xyz` and the package name, `components`. You use `all` component declarations in the package. *Figure 8-4* shows the benefit of having the component declarations in the package; they do not need to be explicitly declared in the architecture. Instead, those declarations occur in the package. It is necessary to compile the package `components` prior to compiling the architecture `STRUCTURAL` since the USE statement has to refer to something that is already in the library. The example illustrates the use of the package.

```
library xyz;                                 Component Source
use xyz.components.all;


architecture STRUCTURAL of COMPARE is        Component Used
signal I: bit;
begin
   U0: XR2 port map (A,B,I);
   U1: INV port map (I, C);
end STRUCTURAL;
```

Figure 8-4

Figure 8-5 shows the package header containing component declarations used above.

```
package components is

component XR2
   port (X, Y: in bit; Z: out bit);
end component;

component INV
   port (X: in bit;  Z: out bit);
end component;

end components;

-----no package body-----------
```

Figure 8-5

Note: The components defined by entity and architecture are not in the package, but separately in library `WORK` or `xyz`.

8.5 Selected Names

VHDL does not permit nested libraries or nested packages. *Figure 8-6* illustrates packages compiled into a library. You use a selected name to get a particular object from a package or from a library. Only three levels of indexing are required to access an item from a package.

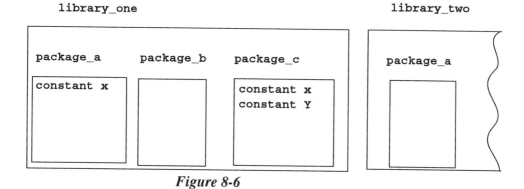

Figure 8-6

The general formats for a selected name are:

`library_name.package_name.item_name` or

`package_name.item_name` or `library.item_name` or

`item_name` if you specify a USE clause

For example, in *Figure 8-6*, to refer to a constant `x` in `package_a`, you can refer to:

`library_one.package_a.x`

To access a procedure `readline` in Package TEXTIO:

`textio.readline (1,a);`

In VHDL 92 hierarchical pathnames are used during simulation to access items. The slash (/) character is used to delineate the crossing of external blocks and the colon (:) is used in crossing internal blocks. Selected name may also be used to indicate part of the pathname.

8.6 USE Statement

To access items in a library, you must use either a selected name or a USE statement. To make library units available for use:

```
USE library_name.{all | item}
```

Examples:

```
USE work.all; use library_one.all;
```

The library name must be previously declared.

Packages hold declarations, which are used in other design units. Items in a package are accessed either by using a selected name or a USE statement. To access a compiled package from a library and make its contents visible with a USE statement, you need to refer to the particular package in a particular library . The syntax is:

> USE   *library_name.package_name.* {all | item}
> USE   *package_name.*{all | item}

Place the USE statement before the entity declaration or architecture declaration. The declarations in the package are then available for your use. You can select all items or a particular item. If you use the `all` option at the end of the USE statement, you can access everything in the package. The alternative is to identify the particular functions or definitions that you want to use. *Figure 8-7* has several different packages; you want to use the `constant x` declared in `package_a` and the `constant Y` out of `package_c`. The USE statement specifies the particular declarations that you want to use from the package.

Example:

```
USE library_one.package_A.x;
USE library_one.package_C.Y;
```

Now `x` and `Y` can be used without selected names. However, to get a constant `x` in `package_c,` you need a selected name.

Note: The scope of a USE statement is only to the design unit it precedes.

```
The default is:

    USE std.standard.all;
```

Therefore, all standard types are available for your use without using selected names (e.g., bit, bit_vector, boolean, and so on).

8.7 General Notes on Procedures in Packages

You can have either WAIT statements or a sensitivity list in the scope of a process, but not both. A process can call a procedure that may also contain a WAIT statement. If so, the calling process cannot have a sensitivity list. The restriction that a process can have a WAIT statement or a sensitivity list includes any function called (see Section 5.11). When documenting your subprograms, you should indicate if you are using WAIT statements because it tells the user not to have a sensitivity list in the calling process. Other aspects that are important to document in the interface to a package are:

- Types of objects (variables, signals, or constants)
- Data types
- Constraints in terms of ranges and values
- What arguments get changed as a result of calling the procedure.

Procedures have arguments that are signals or variables and direction in or out. Packages containing procedures need to document how the procedures are to be called.

8.8 Typical Vendor Packages

When an ASIC vendor or a CAD vendor supplies declarations and definitions to a customer, he usually delivers a VHDL source code package. One of those packages, which is required, is Package STANDARD. BIT_VECTOR, Boolean, and time are data types that are declared in the Package STANDARD. The TEXTIO is also a package provided by the vendor.

In addition, almost all ASIC vendors have extensive component-simulation libraries. They provide VHDL behavioral descriptions, using particular components from their ASIC vendor, so that you can build a structural design.

Some vendors also provide statistical routines that provide random number distribution: functions such as *gaussian*, *uniform*, and *exponential*. These are utility functions that help you do statistical behavioral simulation and stochastic modeling. In addition, some vendors provide a package containing queuing functions to do message passing.

Most vendors distribute packages in VHDL source form. The packages contain type declarations, functions, procedures, etc. Vendors also provide other VHDL source-design units: *entities*, *architectures*, and *packages*. You can compile these and put them into a library. These libraries are referred to as *resource libraries*. The compiled version of these packages can then be accessed with a USE statement.

8.9 IEEE Package 1164

The IEEE standard 1164 defines nine signal strengths within a VHDL package. These nine values are more useful for simulation and synthesis than type bit. The nine values are:

```
U              uninitialized
X              forcing an unknown
0              forcing 0
1              forcing 1
Z              high impedance          Three State
W              weak unknown
L              weak 0
H              weak 1
-              don't care
```

Figure 2-6

The character literal should be in upper case letters and in single quote marks. For example:

    'H'   *NOT*   'h'

Uninitialized is the left most, lowest element of this type; so the default initial value for signals (and variables) of this type is U. Behavioral models may produce an unknown value (X) as an output during operations that are ambiguous -- consider a model of an AND gate that might have an output settling time of 2 ns. During the first 2 ns after an input change, the output could be designated as unknown. Don't care (-) is generally a design specification used to give a logic synthesis tool flexibility in developing the logic circuitry.

The types are:

    std_logic
    std_logic_vector
    std_ulogic
    std_ulogic_vector

To use this package:

```
library IEEE;
use IEEE.STD_LOGIC_1164.all;
```

See Chapter 9 for a further discussion of resolution functions associated with these types.

Package Declaration and Package Body

---

package_declaration

```
package name is
   [subprogram]
   [type]
   [constant]
   [signal]
   [file]
   [alias]
   [USE clause]
   [declarations]
   end [name];
```

---

package_body

```
package_body name is
   [subprogram]
   [type]
   [constant]
   [signal]
   [declarations]
   end [name];
```

- **Summary**

1. The package design unit consists of *package header* and optional *package body*. (Section 8.0)

2. A *package* collects frequently used declarations and stores them in a library. You can then call out a group of declarations without having to repeat them with a USE statement. (Section 8.0)

3. If you do change the package header, you need to recompile anything that is referring to it. (Section 8.1)

4. Rather than having the component declarations in line in a structural design, it is more convenient to have the component declarations in a package. (Section 8.4)

5. In addition to the Package STANDARD, there are vendor packages for additional data types, utility function, and stochastic modeling. (Section 8.7)

-   **Lab #8**

Block Statistical Model of Microcomputer - Using Random-Number Procedures from Synopsys Package

A microcomputer is executing software, which allows it to sample an incoming temperature every .1 second (plus or minus a few milliseconds). Simulate a block statistical model of the microprocessor using the distribution package for a gaussian (normal distribution). Model the microcomputer using an entity `microcheap` as follows:

The median time, when it is available to read, is 90 milliseconds. The standard deviation is 10 milliseconds.

1. Test the gaussian distribution package. Write a process **G**, which gets a random real number variable **x** from the Gaussian distribution package:

```
USE synopsys.distributions.all;
```

Convert **x** to an integer variable itime. Use a 1 ns **WAIT** inside an infinite loop. Monitor `itime`, test the design for 10 samples and record below.

2. The temperature data needs to be read by the `microcheap` module out of a FIFO buffer. The FIFO has output signals indicating when data is available `fifo_empty`. The FIFO has a read input signal, which when pulsed by the `microcheap`, indicates that a data word was read from the FIFO on the FIFO `data_output`. This word is then deleted from the FIFO.

The microcomputer, `microcheap`, has an 8-bit interface for integer `data_input` and a read output signal. The microcomputer model should wait a random amount of time, as described above (convert `itime` to type time, `awhile`, and then read data from the FIFO if not empty). Here is a code outline:

```
WAIT for awhile;     ......around 100 milliseconds
if not (empty)
   read data from data_output
WAIT 25ns;
   assert read signal
   WAIT 75ns;
   deassert read signal
else WAIT for 100ns;
```

List your assumptions. What problems occur?

9 Advanced Topics: Adding Apples &Oranges

This chapter discusses some advanced concepts in VHDL. It contains the following sections:

- Overloading

    - Subprogram Overloading
    - Argument Overloading
    - Operator Overloading
    - Overloading Qualified Expressions

- Resolution Functions and Multiple Drivers

    - Three-State Logic
    - Multiple-Signal Drivers
    - Three-State Buffers
    - Resolution Function for Subtype

- Symbolic Attributes

    - Multi-Range Array Attributes
    - Array Length Attributes
    - Range Attributes
    - Type Attributes - Position Function
    - Signal Attributes
    - Stable Attributes

One of the advantages of the VHDL language is that it allows users to create their own data types, as we previously mentioned in the discussion on enumerated and subtypes in Chapter 4. Additionally, you can define new meanings to operators such as "+". For example:

```
"ABC" + "xyz"
```

Question: What could this mean?

Answer:_____

9.1 Overloading

A designer can refine a design from a more abstract data representation (such as real, and floating-point addition) to a particular bit size and arithmetic operation. At some point in the design, it might be advantageous to describe the particular kind of adder (e..g., carry-look-ahead, etc.) by writing a detailed logical description for the + operation. VHDL permits substituting a new definition.

You can define the operators **+**, **-** , **and**, **or**, etc. associated with your own data types.   The operators can use existing symbolics to invoke new functions. This feature makes VHDL language object-oriented, where an object is the data type and the operator, defined for that data type, works appropriately.   When an existing operator or function is given a new or additional meaning, it is called *overloading*. There are a number of ways to overload objects in VHDL: you can overload subprogram names, overload the number of parameters, or overload operators.

Subprogram Overloading

Two functions can have the same name as long as they have different argument types. In *Figure 9-1* there are two functions named **DECREMENT** used in the same design. The functions are declared with a function declaration statement, and each function has the <u>same</u> name. The first function operates on argument **x** of type **integer**. The second function works on argument **x** of type **real**.

```
• Two subprograms can have the same name.
• The types of the parameters distinguish them.


   function DECREMENT (X: integer) return integer;

   function DECREMENT (X: real) return real;
```

```
   variable A,B : integer
   B := DECREMENT (A);                              function call

Will use the first function  because A is of type integer.
```

Figure 9-1

  In the function call above, the assignment statement indicates that **B** gets the **DECREMENT** of **A**. At first glance, it seems that there is no way to tell which function to call, the first function or the second function. VHDL discriminates the function based upon the type of argument.   In this case, **A** is declared as type **integer**. Therefore, the first decrement function is used: the argument **x** of type **integer**. Functions do not need unique names; the types of arguments distinguish which function to call. This example illustrates an overloaded subprogram name.

Argument Overloading

Another example of overloading is with a varying number of arguments. Note the three functions of the same name called `convert_addr`.

```
PACKAGE p_addr_convert IS
FUNCTION convert_addr (a0,  a1 :INTEGER) RETURN INTEGER;
FUNCTION convert_addr (a0,  a1, a2 :INTEGER) RETURN INTEGER;
FUNCTION convert_addr (a0,  a1, a2, a3 :INTEGER) RETURN INTEGER;
END p_addr_convert;
```

Figure 9-2

The first one has 2 arguments, the second one has 3 arguments, and next one has 4 arguments. You can discriminate the function that you want to call, based upon the number of arguments you list. The example only shows the package header; you would define each of the functions in the package body. When you call the `convert_addr`, depending upon the number of address arguments you provide, you get the corresponding function.

Operator Overloading

In VHDL, the name of a function does not have to be a character name, it can be an operator used between operands. For example, you can call a function with the symbols `+`,   `-`, `>` , `<`, etc. *Figure 9-3* is an example of operator overloading. In the example, the `+` is a function call that shows that `z <= x + y`. Note that `x`, `y`, and `z` are type `BIT_VECTOR`. The predefined Package STANDARD does not define `+` for `BIT_VECTOR`. Normally this is an invalid operator. However, the `+` can be a function that you, or a vendor, wrote to define addition on `bit_vectors`.

```
Predefined operators can be overloaded just like subprograms

   +  -  >  <  and  mod  =  &  *  /

Example of bit-vector addition function call:

     z <= x + y;

where x, y, z are type BIT_VECTOR (3 downto 0)
A new function, usually in a package, defines this + operator.
```

Figure 9-3

The example in *Figure 9-4* shows the declaration of the function +. The arguments are of type BIT_VECTOR and a return value BIT_VECTOR. The arguments A, B will be associated with the operands in the calling expression:

   A is associated with the operand to the left of the +
   B is associated with the operand to the right of the +

In *Figure 9-3*, x is associated with A and y is associated with B. The function + is declared as a ripple carry adder, using Boolean logic. You can define your own adder algorithm and control the synthesis of the hardware circuitry by overriding the standard operator.

```
Package ...
function "+" (A, B: BIT_VECTOR (3 downto 0)) return BIT_VECTOR is
variable SUM: BIT_VECTOR (3 downto 0);
variable CARRY: BIT;
begin
CARRY:= '0';
for I in 0 to 3 loop
   SUM (I) := A(I) xor B(I) xor CARRY;
   CARRY:= (A(I) and B(I) or (A(I) and CARRY) or (CARRY and B(I));
   end loop;
   return SUM;
end;

Example limited to 4 bit vectors...could generalize (unconstrained)
```

Figure 9-4

Supplied operators in VHDL work with the standard data types (for example, logic operators with bit and Boolean, and arithmetic with integer and real). The overloading of operators permits you to provide additional functions. These functions can apply to new data types.

   Remember:

   • Supplied operators work only with standard types.

   • Overloading allows definition of operators for any type, including user-defined types.

Another example is the addition of strings. Typically, adding strings is not a valid operation. You can, however, define   "+"   as a function to concatenate two strings together, resulting in one string.

Question: How do you write an addition of strings function? Write it in this box.

Overloading and Qualified Expressions

The choice of function you are calling depends upon the type, and since VHDL provides enumerated types, you sometimes have ambiguities. You may need a qualified expression to qualify a name by giving a type, followed by a tick mark (′), followed by the argument that is being qualified in parentheses (see Section 4.1 on qualified expressions). For example:

```
month'(June)
name'(June)
```

Remember, a qualifier:

- Specifies the type of expression.
- Is useful for overloaded operators.
- Can be used to specify the return type expected from a complex expression.

Overloading is a modern computer-science idea that allows you to create readable code. You can create a natural looking expression involving infix operators which call user-written functions.

9.2 Resolution Functions and Multiple Drivers

VHDL provides a technology-independent and user-definable extension to support hardware concepts, such as bussing, wiring, and three-state operations. Rather than specific built-in operations, the language allows user-written or vendor-written *resolution functions*, which define (resolve) the meaning of shorting or "dotting" signals together. In addition, as an added elegance, these resolution functions are called automatically, the *implicit function call*. This call is made whenever a statement assigns a value using a signal assignment statement to a "resolved bus". Therefore, the hardware-description does not need to contain the function call (or the resolution function itself). The "trick" in VHDL is to associate a resolution function with a particular data type, and any assignment to a signal of that type automatically invokes the particular resolution function. Therefore, you can have a **wired-and** type as well as a **wired-or** type or **three-state** bus.

In VHDL, a signal has a driver and that driver determines the values on the signal. However, just as you can have hardware busses with multiple sources, you can have multiple signal drivers. Evaluate the example below. The example says that **c** gets **A** while **c** gets **B**. You have a contradiction of multiple simultaneous assignments to the same signals. You need a resolution function to determine meaning (wired-and, wired-or, etc.).

```
C <= A;
C <= B;
```

This example is ambiguous because it wires both **A**, **B** to **c**; it needs to be resolved . VHDL supports multiple drivers of a signal and a resolution of the meaning (wired-or, wired-and, three state) is required.

Three-State Logic

The next example has a signal of type **BIT3**, which has the **z** state, high impedance not driving. Since the signal is three-state, it is valid to have more than one driver (one driver is three-stated or not really driving the bus).

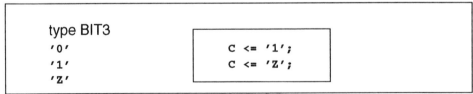

Figure 9-5

Question: How does a simulator resolve the logic values?
Answer:

Multiple Signal Drivers

In VHDL, you should have only one active driver for a signal per process (if it is being driven for that process). In the following example, there is a signal **s** that is being driven by two processes, **process a** and **process b**.

```
architecture pack_invalid_example is

    signal S: BIT3;                    b: process
    a: process....                     begin
    begin                              S <= 'z';
    S <= '1';                          end process;
    end process;
```

• VHDL requires each bus driver to be in a separate process.
• Designer can specify wired-and, wired-or, or three-state or any other meaning using a type that has a resolution function.

Figure 9-6

Notice that in **process a**, the signal is being driven to a **1** and in **process b**, it is driven to **z**. This example looks like a valid expression. It has two drivers, each in separate processes. The example, however, does require a resolution function for **BIT3** to indicate the meaning when you drive a **1** and a **z**. Since **BIT3** was declared without a resolution function, it is an error to have multiple drivers as shown. Note that it does not have a predefined meaning. You need to have a resolution function that is automatically called when you drive bus **s**. The resolution function needs to return a value that sets **s** to some value (in this case, '0', '1', or 'z'). The language is extensible so that it can handle all kinds of logic systems, as long as you have resolution functions that define the meaning. Resolution functions of this type are usually defined by vendors (for example, if they have three-state buffers).

By associating a particular type to a resolution function, you know that the signal is a wired-and type or wired-or type, etc. For example:

```
BIT3_and
BIT3_OR
```

could be resolved types for BIT3.

Note: If **BIT3** is defined without a resolution function as part of the type declaration, you get an error message during analysis or simulation.

Three-State Buffers

Below is an example of three-state buffers.

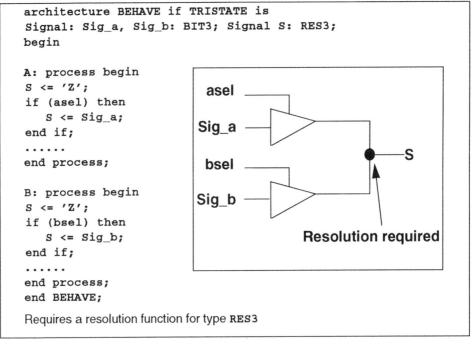

```
architecture BEHAVE if TRISTATE is
Signal: Sig_a, Sig_b: BIT3; Signal S: RES3;
begin

A: process begin
S <= 'Z';
if (asel) then
    S <= Sig_a;
end if;
......
end process;

B: process begin
S <= 'Z';
if (bsel) then
    S <= Sig_b;
end if;
......
end process;
end BEHAVE;
```

Requires a resolution function for type RES3

Figure 9-7

In the hardware schematic, there are two three-state drivers that drive signal **s**. Each driver has a select line, **asel** and **bsel**. Only one of them should be active at a time. When one is active, all others should be disabled. In *Figure 9-7*, you have process **A** and process **B**, each of which initializes **s** to the 'z' state ("not-driven state"). Signal **s** is of type **RES3** and *has* a resolution function (not shown here). Depending on the select line, **asel** or **bsel**, if one of them is true, a value is driven on signal **s**. Only process **A** or process **B** should drive one of the signals out to **s**. The other one (the inactive process) has the select line set low, which causes one process to drive a bit out to **s** (perhaps a 1 or a 0) and the other process to drive a bit out to **z**. The resolution function associated with **RES3** sees two values being driven and determines if you have **z** and a **1**, or a **z** and a **0**; and as a result, returns the value 1 or a 0.

A resolution function is a user-written or vendor-provided function that resolves the value of a bus or a wire when there is more than one driver. It is important to notice that in the example shown in *Figure 9-7*, there is no explicit call of the resolution function. This example calls a function during simulation. Because there is more than one driver, the system calls the function automatically, whenever a value is driven. A call is made to a function associated with a particular data type **RES3**, and that function is passed all of the values being driven.

The argument to the resolution function is an array of all the values being driven. In the case of example in *Figure 9-7*, it is an array of two values: the value from process **A** and the value from process **B**. The resolution function resolves, based on a table of values, and returns one value. All of the arguments are passed automatically, independent of the number of drivers that undergo a signal change. Any time a signal changes, the function is called automatically and the resolution function resolves the values.

A function:

- Returns one value.

- Has all arguments of mode input.

- Has all arguments passed by value.

A resolution function:

- Resolves the value of a signal with multiple drivers.

- Is required when a signal has multiple drivers.

- Is called when any driver has an event occur.

- Is passed a variable-length array of signal values to be resolved.

- Is a user-definable function: typically wired-and or three-state.

- Associates with a subtype (which becomes a resolved type).

Since every assignment to a resolved bus requires a function call, the simulator execution is slower than the execution of a nonresolved bus.

Resolution Function for Subtype

A resolution function is associated with a subtype. Any signal that needs to be resolved is declared of that subtype, which is how a resolution function is tied to a subtype. The syntax is:

Subtype *resolve_type_name* **is** *resolution_function base_type*

For example :

```
TYPE BIT3 IS ('0', '1', 'Z');
TYPE bit3_vector IS ARRAY (natural range<>) OF BIT3;

FUNCTION resolve (s:bit3_vector) RETURN BIT3;      Resolution Function
   .........                                       Declaration
   ........
SUBTYPE res3 IS resolve BIT3;                      Resolution Function
                                                   Use

```

Figure 9-8

The subtype declaration **res3** is the resolved type for **BIT3**. It names a new type, in this case the type **res3**. It also names the resolution function **resolve** and the type (in this case, the type of **BIT3**) that is going to be resolved. The example resolution function **resolve** declares an array and a type **bit3_vector**, which is an array of **BIT3**. There is a function named **resolve** that brings in argument **s** of type **bit3_vector** and returns a value of type **BIT3**. The array **s** brings in the the various drivers.

The resolution function **resolve** is not shown here but would loop over the arguments and calculate a return value. Typically, a CASE statement or a table look-up utility with a constant array is used. See Section 9.3 for processing variable-length arrays using the **'length** attribute.

Typically, a VHDL user does not write resolution functions but is a user of predefined resolution functions. Resolution functions resolve the subtype of data, so you must choose which subtype of data to use. For example:

Resolution Functions in IEEE Package 1164

std_logic and std_logic_vector are resolved types and you can have multiple drivers for signals of this type. std_ulogic and std_ulogic_vector are unresolved types and you cannot have more than one driver.

9.3 Symbolic Attributes

VHDL has symbolic attributes that allow you to write more generalized code. Instead of using constants or literals in code, you can use attributes that are understood and known at compile time. Some of these attributes are predefined in VHDL, others are provided by a CAD vendor. Also, you can add your own defined attributes (see Chapter 10 for synthesis attributes). There are attributes that relate to arrays, types, ranges, position, and signal characteristics.

The attributes below work with arrays and types.

- **aname** `'LEFT` returns left bound of index range.

- **aname** `'RIGHT` returns right bound of index range.

- **aname** `'HIGH` returns upper bound of index range.

- **aname** `'LOW` returnns lower bound of index range.

' designates a separator between the name (in this case, an array name) and the attribute. This ' works with array types and arrays themselves. Below is an example.

```
signal my_array: bit_vector (31 downto 0);
```

my_array is an array **31 downto 0** and rather than using the numbers 31 and 0 in your code, you can refer to **my_array'left** and **my_array'right**. Rather than using the high and low numbers you can use **my_array'high** and **my_array'low** in the example below.

```
TYPE t_ram_data IS ARRAY (0 to 511) OF INTEGER;
VARIABLE ram_data: t_ram_data;
.

.
FOR i IN ram_data'LOW TO ram_data'HIGH LOOP
   ram_data(i) := 0;
END LOOP;
```

Figure 9-9

The example in *Figure 9-9* contains a loop for **i** in **ram_data'low** to **ram_data 'high**. Since **ram_data** is described as a type whose subscripts range from 0 to 511, rather than using the constants 0 and 511 in the **FOR LOOP**, you can use the symbolic constants **ram_data'low** and **ram_data'high**. There is no difference in the constants from the symbolic attributes. If sometime in the future, however, you changed the numbers 0 and 511, you would only have to change them in the type declaration and the **FOR LOOP** would be adjusted automatically.

Multi-Range Array Attributes

With multi-dimensional arrays, you can specify the attributes for each of the ranges.

```
                        •   'LEFT (n) returns left bound of index range n.

                        •   'RIGHT (n)  returns right bound of index range n.

                        •   'HIGH (n) returns upper bound of index range n.

                        •   'LOW (n) returns lower bound of index range n.

Example:      n =     1      2

Variable: memory (0 to 5, 0 to 7) of mem_data;
                                   ▲
                                   |
                     memory 'RIGHT (2)
```

Figure 9-10

Using multirange attributes, you put the index range, the first range, second range, or third range (whatever number it is) in parentheses. The example shows **memory 'right(2)** as the second range subscript range (0 to 7) and rightmost value is 7.

Array Length Attributes

Another example of symbolic attributes is the length attribute.

```
'LENGTH returns length of array or array type

Example:
PROCESS (a)
   TYPE bit4 IS ARRAY (0 to 3) OF BIT;
   TYPE bit_strange IS ARRAY (10 to 20) OF BIT;
   VARIABLE len1, len2: INTEGER;
BEGIN
   len1 := bit4'LENGTH;                    returns 4
   len2 := bit_strange'LENGTH;             returns 11
END PROCESS;
```

Figure 9-11

Rather than specifying the length of an array with a numeric value, you can do it symbolically. For example, `bit4'length` returns 4. It is the same for `bit_strange'length`, which returns 11. In this example, you can see that `10 to 20` are the range of subscripts of that array and the length is 11.

Range Attributes

You can use a symbolic value as an actual range, for example:

```
FOR i IN STUFF'RANGE LOOP
```

```
name 'RANGE returns the range of a particular type.

name 'REVERSE_RANGE returns the range of a particular type in reverse
order.

Example:

FUNCTION vector_to_int (stuff: bit_vector) RETURN INTEGER IS
   VARIABLE result : INTEGER := 0;
BEGIN
   FOR i IN STUFF'RANGE LOOP
   . . . . . . . . . .

```

Figure 9-12

The range attribute returns the range of an object. You can use the name `'RANGE` and the name `'REVERSE_RANGE`. In the example, the function is written for the argument `STUFF` of type `bit_vector`; the length is not given. When the function is invoked on a particular array, then a range value is passed in and the loop can handle any length vector. The best use of attributes is when you do not actually know the length of an array and you want to provide for varying sizes; when you call the function, the argument has a fixed length.

Type Attributes - Position Functions

Another example of symbolic attributes is for enumerated types. Enumerated types has the notion of SUCCessor and PREdecessor, left or right of the position number of the value:

- Typename 'SUCC(v) returns next value in type after **v**.

- Typename 'PRED(v) returns previous value in type before **v**.

- Typename 'LEFTOF(v) returns value immediately to left of **v**.

- Typename 'RIGHTOF(v) returns value immediately to right of **v**.

- Typename 'POS(v) returns type position number of **v**.

- Typename 'VAL(p) returns type value from **p**.

- Typename 'BASE returns base type of type or subtype.

These functions are used for enumerated types and can be used for symbolic code. Look at the example below.

```
TYPE color IS (red, blue, green, yellow, brown, black);
SUBTYPE color_gun IS color RANGE red TO green;
VARIABLE a: color;

a := color'LOW;
a := color'SUCC (red);
a := color_gun'BASE'RIGHT;
a := color'BASE'LEFT;
a := color_gun'BASE'SUCC (green);
```

Figure 9-13

You have:

color   'LOW is red

color 'SUCC (red) gives blue

color_gun 'BASE'RIGHT gives black

color 'BASE'LEFT equals red

color_gun 'BASE'SUCC (green) gives yellow

Signal Attributes

There is another class called *signal attributes* that work on signals. Signal attributes tell us something about simulation time events:

* **signame** `'EVENT` returns true if an event occurred this time step.

* **signame** `'ACTIVE` returns true if a transaction occurred this time step.

* **signame** `'LAST_EVENT` returns the elapsed time since the previous event transaction.

* **signame** `'LAST_VALUE` returns previous value of signal before the last event transition.

* **signame** `'LAST_ACTIVE` returns time elapsed since the previous transaction occurred.

Signal attributes allow you to do some complicated tests. For example:

```
ENTITY dflop IS
  PORT (d, clk: IN mvl7; q: OUT mvl7);
END dflop;

ARCHITECTURE dff OF dflop IS
BEGIN
  PROCESS (clk)
  BEGIN
    IF (clk = '1') and (clk'EVENT)
        and (clk'LAST_VALUE= '0') THEN q <= d;
    END IF;
  END PROCESS;
```

Figure 9-14

The process tests if `clk` is a 1 and `clk'event`, which means the clock is changed to a 1. If the last previous value of `clk` is a zero, then you have a true rising edge. `clk` was zero and is now a 1. The example qualifies a situation based on the current and previous values. You can also test for setup time, etc.

Question: Why test previous value for 0?

Answer:

You can also have signals that are derived from other signals. You can have a signal that delays a signal, using the delayed operator of time. You can also create some signals based on whether a signal has been stable for a certain amount of time or quiet for a certain amount of time.

- **Summary**

---

1. Overloading operators creates more readable code. (Section 9.1)

2. Qualified expressions can define ambiguous types. (Section 9.1)

3. *Resolution functions* resolve the value of a signal with multiple drivers. It associates with a subtype, which uniquely identifies a bus or a wire. (Section 9.2)

4. *Predefined attributes* return information about objects. They are useful for performing certain types of functions such as: timing checks, bounds checking, clock-edge detection, and type conversion. (Section 9.3)

5. User-defined attributes can be used to attach data to VHDL objects. (Section 9.3)

-   Answers to Questions

Question: What could this mean?

Answer:_____

Question: How do you write an addition of strings function? Write it in this box.
Answer:
```
Function "+" (a, b: string) return string is
return (a & b);
end
```

Although Chapter 2 emphasizes not mixing types in expressions, user packages can define mixed-mode operations.
For example:

Question: How does a simulator resolve the logic values?

Answer: Type `bit3` is a resolved type and has a function that resolves the 9 combinations:
      0,0   0,1   0,Z   1,0   1,1   1,Z   Z,0   Z,1   Z,Z
Depending on how the function is written, it could be `wired-or`, `wired-and`, `three-state`, etc.

Question: Why test previous value for 0?

Answer: To distinguish it from a transition from an unknown.

10 | VHDL &
Logic Synthesis

This chapter discusses how VHDL is used for logic synthesis. It contains the following sections:

- Synthesis-Ready Code

- CASE Statement Synthesis

- FOR Statement Synthesis

- A 4-Bit Adder

- Synthesis and the WAIT Statement

- State Machines in VHDL

- Predefined Attributes for Synthesis

Although VHDL was created as a documentation language and later as a simulation language, it can be used for the technology-independent description of a logic chip. A compiler can process certain VHDL statements and synthesize logic-gate netlists and schematics for ASIC libraries or synthetic libraries. To synthesize VHDL statements, the compiler requires that the VHDL language be written at a restricted level or RTL (register transfer) level, which means there are certain behavioral statements that are not meaningful or are not synthesizable. CAD vendors differ in their synthesizable subsets of VHDL; some language changes have been proposed to support logic synthesis.

In the following sections some examples are shown for logic synthesis from VHDL as supported by the Synopsys VHDL Compiler, Version 2. Additionally, some references are made to a Synopsys Package, which declares new data types *signed* and *unsigned*, and overloaded operators.

10.1 Synthesis-Ready Code

An example of a statement that is not ready for synthesis is shown below.

```
    signal x : integer;
    process
    begin
    x <= x + 1 after 10 ns;
```

Figure 10-1

The example has **x** gets **x + 1 after 10 ns**. Some synthesizers may not be able to build logic from this code, and so, you need another way of specification. Some parts of the language are not synthesizable, but generally many of the statements in the language *are* synthesizable.

Below is an example of synthesis of the logical statement:

```
    x <= (a and b) or (c and (or e not f)) xor g;
```

This VHDL statement can be translated automatically into logic gates, and then optimized for a particular set of components from a vendor library for speed and area .

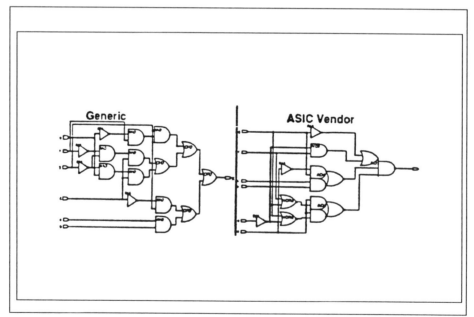

Figure 10-2

10.2 CASE Statement Synthesis

Another example of synthesis is when you go   from a CASE statement to a
mulltiplexer. In some cases, the compiler produces a schematic and illustrates a
4-to-1 multiplexer.

```
use work.synopsys.all
entity VHDL is
port (A, B, C, D : In BIT;
       CHOICE   : In ENUM;
       Z : out BIT);
end VHDL;


architecture VHDL_1 of VHDL is
begin
process (CHOICE) begin
   case CHOICE is
      when USE_A => Z <= A;
      when USE_B => Z <= B;
      when USE_C => Z <= C;
      when USE_D => Z <= D;
   end case;
   end process;
end VHDL_1;
```

Figure 10-3

10.3 FOR Statement Synthesis

The example below illustrates the synthesis of a FOR loop. The FOR loop builds a cascading array using EXCLUSIVE OR gates.

```
entity VHDL is
   port (WORD : in BIT_VECTOR (0 to 7);
      PARITY : out BIT);
end VHDL;

architecture VHDL_1 of VHDL is
begin
   process (WORD)
      variable ODD : bit;
   begin
      ODD := '0';
   for i in 0 to 7 loop
      ODD := ODD xor WORD (i);
   end loop;
   PARITY <= ODD;
   end process;
end VHDL_1;
```

Figure 10-4

10.4 A 4-Bit Adder

VHDL can also synthesize infix operators such as + and - and *. Going from the statement C <= A + B, depending upon the size of A and B, the compiler generates an appropriate adder for A + B. In this case, 4 bits of A and 4 bits of B.

```
entity VHDL is
    port (A, B : in INTEGER range 0 to 15;- Implies a 4-bit Result
          C : out INTEGER range 0 to 15);
end VHDL;

architecture VHDL_1 of VHDL is
begin
    C <= A + B;
end VHDL_1;
```

Figure 10-5

10.5 Synthesis and the WAIT Statement

Some of the differences between synthesis and the full language of VHDL have to
do with the treatment of the WAIT statement. One synthesis technique is to use the
WAIT statement to represent a clock and to specify a type of clock.

```
LIBRARY IEEE;
use IEEE.STD_LOGIC_1164.all;
use IEEE.STD_LOGIC_UNSIGNED.all;
entity COUNTER4 is
   port (UPDOWN : in bit;
      CLK : in bit;
      DATAOUT : out std_logic_vector (3 downto 0));
end COUNTER4;

architecture BEHAVIOR of COUNTER4 is
   begin
      process

   begin

Wait until CLK'event and CLK = '1';

   if (UPDOWN = '1') then
         DATAOUT <= DATAOUT + "0001";
      else
         DATAOUT <= DATAOUT - "0001";
      end if;
   end process;
end BEHAVIOR;
```

Figure 10-6

The example above says wait until a clock has an event, which means a change,
and goes to a **1**; this finds the next rising edge of the **CLK**. A counter is built, in
this case, an updown counter. It has a 4-bit count value and 4 flip-flops are
inferred.

In the example above, **DATAOUT** is declared to be of type std_logic_vector. This
data type is declared in package **IEEE.STD_LOGIC_1164**, which is made visible by
the USE statement. The package **IEEE.STD_LOGIC.UNSIGNED** has overloaded
arithmetic operators, as shown above (**+** and **-**).

The next example shows a single flip-flop being generated for this WAIT statement. There is a clock signal called CLOCK that is in the WAIT statement and a toggle flip-flop with an ENABLE. Note that the hardware synthesized matches the statements in the VHDL code.

```
entity VHDL is
    port (ENABLE : in BIT;
        CLOCK : in BIT;
        TOGGLE : buffer BIT);
end VHDL;

architecture VHDL_1 of VHDL is
begin
    process begin
        wait until CLOCK'event and CLOCK = '1';
        if (ENABLE = '1') then
            TOGGLE <= not TOGGLE;
        end if;
    end process;
end VHDL_1;
```

Figure 10-7

10.6 State Machines in VHDL

Another example of VHDL synthesis is state machines. State machines are a popular design technique.The design example   below uses the vending drink machine control unit first shown in Chapter 3. This example is the state-machine version.

```
entity DRINK_STATE_VHDL is
  port(NICKEL_IN, DIME_IN, QUARTER_IN, RESET: in BOOLEAN;
       CLK: INBIT;
       NICKEL_OUT, DIME_OUT, DISPENSE: out BOOLEAN);
end;

architecture BEHAVIOR of DRINK_STATE_VHDL is
  type STATE_TYPE is (IDLE, FIVE, TEN, FIFTEEN,
                      TWENTY, TWENTY_FIVE, THIRTY,
                      OWE_DIME);
  signal CURRENT_STATE, NEXT_STATE: STATE_TYPE;
begin

  process(NICKEL_IN, DIME_IN, QUARTER_IN,
          CURRENT_STATE, RESET, CLK)
  begin
    -- Default assignments
    NEXT_STATE <= CURRENT_STATE;
    NICKEL_OUT <= FALSE;
    DIME_OUT <= FALSE;
    DISPENSE <= FALSE;

    -- Synchronous reset
    if(RESET) then
      NEXT_STATE <= IDLE;
    else

      -- State transitions and output logic
case CURRENT_STATE is
   when IDLE =>
       if (NICKEL_IN) then
          NEXT_STATE <= FIVE;
       elsif (DIME_IN) then
          NEXT_STATE <= TEN;
       elsif (QUARTER_IN) then
          NEXT_STATE <= TWENTY_FIVE;
       end if;
```

Figure 10-8

```
   when FIVE =>
      if (NICKEL_IN) then
         NEXT_STATE <= TEN;
      elsif (DIME_IN) then
         NEXT_STATE <= FIFTEEN;
      elsif (QUARTER_IN) then
         NEXT_STATE <= THIRTY;
      end if;
   when TEN =>
      if (NICKEL_IN) then
         NEXT_STATE <= FIFTEEN;
      elsif (DIME_IN) then
         NEXT_STATE <= TWENTY;
      elsif (QUARTER_IN) then
         NEXT_STATE <= IDLE;
         DISPENSE <= TRUE;
      end if;
   when FIFTEEN =>
      if (NICKEL_IN) then
         NEXT_STATE <= TWENTY;
      elsif (DIME_IN) then
         NEXT_STATE <= TWENTY_FIVE;
      elsif (QUARTER_IN) then
         NEXT_STATE <= IDLE;
         DISPENSE <= TRUE;
         NICKEL_OUT <= TRUE;
      end if;
when TWENTY =>
   if(NICKEL_IN) then
      NEXT_STATE <= TWENTY_FIVE;
   elsif(DIME_IN) then
      NEXT_STATE <= THIRTY;
   elsif(QUARTER_IN) then
      NEXT_STATE <= IDLE;
      DISPENSE <= TRUE;
      DIME_OUT <= TRUE;
   end if;

when TWENTY_FIVE =>
   if(NICKEL_IN) then
      NEXT_STATE <= THIRTY;
   elsif(DIME_IN) then
      NEXT_STATE <= IDLE;
      DISPENSE <= TRUE;
   elsif(QUARTER_IN) then
      NEXT_STATE <= IDLE;
      DISPENSE <= TRUE;
      DIME_OUT <= TRUE;
      NICKEL_OUT <= TRUE;
   end if;
```

Figure 10-8 Continued

```
   when THIRTY =>
     if(NICKEL_IN) then
        NEXT_STATE <= IDLE;
        DISPENSE <= TRUE;
     elsif(DIME_IN) then
        NEXT_STATE <= IDLE;
        DISPENSE <= TRUE;
        NICKEL_OUT <= TRUE;
     elsif(QUARTER_IN) then
        NEXT_STATE <= OWE _DIME;
        DISPENSE <= TRUE;
        DIME_OUT <= TRUE;
     end if;

 when OWE_DIME =>
        NEXT_STATE <= IDLE;
        DIME_OUT <= TRUE;

 end case;
 end if;
 end process;


 -- for flip/flops

 process
 begin
    wait until CLK'event and CLK = '1';
    CURRENT_STATE <= NEXT_STATE;
 end process;

 end BEHAVIOR;
```

Figure 10-8 Continued

10.7 Predefined Attributes for Synthesis

VHDL provides predefined attributes, which in this case are predefined by the CAD vendor. You can automatically generate different versions of the circuitry by setting these synthesis attributes. Some typical vendor's attributes are shown below for area, speed, and encoding.

Design Constraints

Synthesis constraints for designs are defined to control the target chip die area and chip operating frequency.

```
attribute MAX_AREA :        real;
  -- Maximum desired area, in area units.

attribute MAX_TRANSITION : real;
  -- Maximum allowable transition time

attribute MAX_DELAY :        real;
        -- Maximum allowable delay time, from any
        -- input signal connected to the output

attribute LOAD : real;
  -- Loading on output port, in library load units.

attribute ARRIVAL : real;
  -- Expected input signal arrival time

attribute DRIVE_STRENGTH : real;
  -- Input signal's drive strength
```

Enumeration Encoding

Enumeration types are coded by default - the first enumeration literal is assigned the value 0, the next enumeration literal is assigned the value 1, etc.
The ENUM_ENCODING attribute allows you to specify numeric code values for the enumerated type implementation.

```
attribute ENUM_ENCODING: STRING;

type COLOR is (RED, GREEN, YELLOW, BLUE, VIOLET);
attribute ENUM_ENCODING of
  COLOR: type is "010 000 011 100 001";
```

All of the above examples are CAD-vendor specific.

- **Summary**

---

1. To synthesize VHDL statements, the compiler requires that the VHDL language be written at the *register-transfer level*, which means there are certain behavioral statements that are not meaningful or are not synthesizable. (Section 10.0)

2. Some parts of the VHDL language are not *synthesizable*, but many of the statements are synthesizable. (Section (10.1)

3. You can automatically generate different versions of the circuitry in terms of area and speed, depending upon predefined attributes. (Section 10.7)

A Reserved Words

The identifiers listed below are called reserved words and are reserved for significance in the language. Reserved words differing only in the use of corresponding upper and lower case letters are considered as being the same. The reserved word range is also used as the name of a pre-defined attribute. A reserved word must not be used as an explicitly declared identifer. The words shown in light font are new in VHDL 92.

abs	element	nand	select
access	else	new	severity
after	elsif	next	signal
alias	end	nor	shared
all	entity	not	sla
allow	exit	null	sll
and			sra
architecture	file	of	srl
array	for	on	subtype
assert	function	open	
attribute		or	
	generate	others	then
begin	generic	out	to
block	group		transport
body	guarded	package	type
buffer		port	
bus	if	postponed	unaffected
	impure	private	units
case	in	procedure	until
component	initial	process	use
configuration	inout	pure	
constant	is		variable
		range	
disconnect	label	record	wait
downto	library	register	when
	linkage	reject	while
	literal	rem	with
	loop	report	
		return	xnor
	map	rol	xor
	mod	ror	

B Application Examples

This appendix contains VHDL example code. It contains three examples:

- Vending Drink Machine - Count Nickels
- Structural Description of a 3-Bit Counter
- Carry-Lookahead Adder

B.1 Vending Drink Machine - Count Nickels

The first application example is a control unit for a vending drink machine. The circuit reads signals from a coin input unit and sends outputs to a change dispensing unit and a drink dispensing unit. The examples assume that there is only one kind of soft drink dispensed. It is a clocked design with CLK and RESET input signals. The price of the drink is 35 cents. Input signals from the coin input unit are NICKEL_IN (nickel deposited), DIME_IN (dime deposited), and QUARTER_IN (quarter deposited). Output signals to the change dispensing unit are NICKEL_OUT and DIME_OUT. The output signal to the drink dispensing unit is DISPENSE (dispense drink). This VHDL description counts nickels. A counter counts the number of nickels deposited. This counter is incremented by 1 if the deposit is a nickel, by 2 if it is a dime, and by 5 if it is a quarter.

```
entity DRINK_COUNT_VHDL is
  port(NICKEL_IN, DIME_IN, QUARTER_IN, RESET: in BOOLEAN;
       CLK: in BIT;
       NICKEL_OUT, DIME_OUT, DISPENSE: in BOOLEAN);
end;

architecture BEHAVIOR of DRINK_COUNT_VHDL is
  signal CURRENT_NICKEL_COUNT,
         NEXT_NICKEL_COUNT: INTEGER range 0 to 7;
  signal CURRENT_RETURN_CHANGE, NEXT_RETURN_CHANGE : BOOLEAN;
begin
  process(NICKEL_IN, DIME_IN, QUARTER_IN, RESET, CLK,
          CURRENT_NICKEL_COUNT, CURRENT_RETURN_CHANGE)
    variable TEMP_NICKEL_COUNT: INTEGER range 0 to 12;
  begin
    -- Default assignments
    NICKEL_OUT <= FALSE;
    DIME_OUT <= FALSE;
    DISPENSE <= FALSE;
    NEXT_NICKEL_COUNT <= 0;
    NEXT_RETURN_CHANGE <= FALSE;

    -- Synchronous reset
                if (not RESET) then
                    TEMP_NICKEL_COUNT := CURRENT_NICKEL_COUNT;
```

Figure B-1

```
-- Check whether money has come in
     if (NICKEL_IN) then
       -- NOTE:  This design will be flattened, so
       --    these multiple adders will be optimized
       TEMP_NICKEL_COUNT := TEMP_NICKEL_COUNT + 1;
     elsif(DIME_IN) then
       TEMP_NICKEL_COUNT := TEMP_NICKEL_COUNT + 2;
     elsif(QUARTER_IN) then
       TEMP_NICKEL_COUNT := TEMP_NICKEL_COUNT + 5;
     end if;

-- Enough deposited so far?
     if(TEMP_NICKEL_COUNT >= 7) then
       TEMP_NICKEL_COUNT := TEMP_NICKEL_COUNT - 7;
       DISPENSE <= TRUE;
     end if;

     -- Return change
     if(TEMP_NICKEL_COUNT >= 1 or
       CURRENT_RETURN_CHANGE) then
       if(TEMP_NICKEL_COUNT >= 2) then
         DIME_OUT <= TRUE;
         TEMP_NICKEL_COUNT := TEMP_NICKEL_COUNT - 2;
         NEXT_RETURN_CHANGE <= TRUE;
       end if;
       if(TEMP_NICKEL_COUNT = 1) then
         NICKEL_OUT <= TRUE;
         TEMP_NICKEL_COUNT := TEMP_NICKEL_COUNT - 1;
       end if;
     end if;

     NEXT_NICKEL_COUNT <= TEMP_NICKEL_COUNT;
   end if;
 end process;

 -- Remember the return-change flag and
 -- the nickel count for the next cycle
 process
 begin
   wait until CLK'event and CLK = '1';
   CURRENT_RETURN_CHANGE <= NEXT_RETURN_CHANGE;
   CURRENT_NICKEL_COUNT <= NEXT_NICKEL_COUNT;
 end process;

end BEHAVIOR;
```

Continue Figure B-1

B.2 Structural Description of a Design Entity

The second example shows a structural (netlist) description of a design entity called
COUNTER3.

```
architecture STRUCTURE of COUNTER3 is
    component DFF
        port(CLK, DATA: in BIT;
        Q: out BIT);
    end component;

    component AND2
        port (I1, I2: in BIT);
        O: out BIT);
    end component;

    conponent OR2
        port (I1, I2: in BIT;
        O: out BIT);
    end component;

    component NAND2
        port (I1, I2: in BIT;
        O: out BIT);
    end component;

    component XNOR2
        port (I1, I2: in BIT;
        O: out BIT);
    end component;

    component INV
        port (I: in BIT;
        O: out BIT);
    end component;

signal N1, N2, N3, N4, N5, N6, N7, N8, N9: BIT;
begin
    u1: DFF port map (CLK, N1, N2);
    u2: DFF port map (CLK, N5, N3);
    u3: DFF port map (CLK, N9, N4);
    u4: INV port map (N2, N1);
    u5: OR2 port map (N3, N1, N6);
    u6: NAND2 port map (N1, N3, N7);
    u7: NAND2 port map (N6, N7, N5);
    u8: XNOR2 port map (N8, N4, N9);
    u9: NAND2 port map (N2, N3, N8);
    COUNT (0) <= N2;
    COUNT (1) <= N3;
    COUNT (2) <= N4;
end STRUCTURE;
```

Figure B-2

B.3 Carry-Look-Ahead Adder

The third example demonstrates the use of concurrent procedure calls to build a 32-bit carry-lookahead adder. The adder is built by partitioning the 32-bit input into eight slices of four-bits each. Each of the eight slices computes *propagate* and *generate* values using the PG procedure. *Figure B.3* shows the overall structure.

Propagate (output P from PG) is '1' for a bit position if that position propagates a carry from the next lower position to the next higher position. Generate (output G) is '1' for a bit position if that position generates a carry to the next higher position, regardless of the carry-in from the next lower position.

The carry-lookahead logic reads the carry-in, propagate, and generate information computed from the inputs. It computes the carry value for each bit position. This logic makes the addition operation just an XOR of the inputs and the carry values. The carry values are computed by a three-level tree of four-bit carry-lookahead blocks:

The first level of the tree computes the 32 carry values and the eight group propagate and generate values. Each of the first-level group propagate and generate values tell if that 4-bit slice propagates and generates carry values from the next lower group to the next higher. The first-level lookahead blocks read the group carry computed at the second level.

The second-level lookahead blocks read the group propagate and generate information from the four first-level blocks, then compute their own group propagate and generate information. They also read group carry information computed at the third level to compute the carries for each of the third-level blocks.

The third-level block reads the propagate and generate information of the second level to compute a propagate and generate value for the entire adder. It also reads the external carry to compute each second-level carry. The carry-out for the adder is '1' if the third-level generate is '1', or if the third-level propagate is '1' and the external carry is '1'.

The third-level carry-lookahead block is capable of processing four second-level blocks. Since there are only two, the high-order two bits of the computed carry are ignored, the high-order two bits of the generate input to the third-level are set to zero "00", and the propagate high-order bits are set to "11", which causes the unused portion to propagate carries but not to generate them.

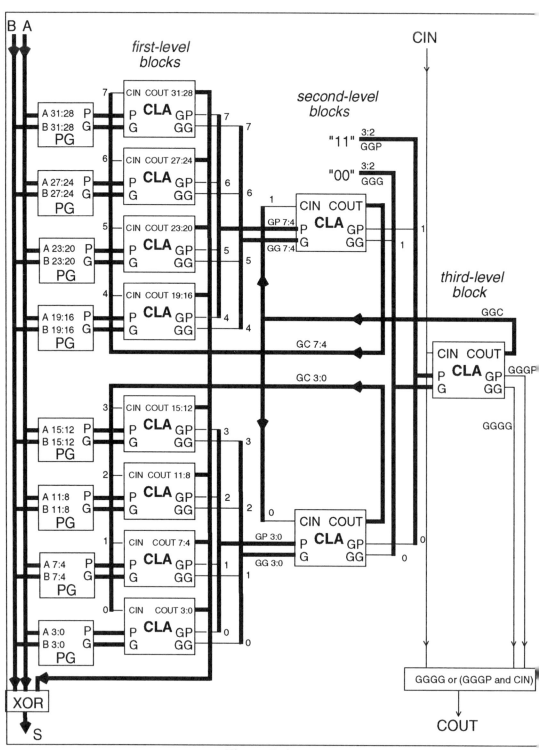

Figure B-3

The VHDL implementation of the design in *Figure B-3* is done with four procedures:

- **CLA**, which a four-bit carry-lookahead block.

- **PG**, which computes first-level propagate and generate information.

- **SUM**, which computes the sum by **XOR**ing the inputs with the carry values computed by **CLA**.

- **BITSLICE**, which collects together the first-level **CLA** blocks, the **PG** computations and the **SUM**. This procedure performs all the work for a four-bit value except for the second- and third-level lookaheads.

Figure B-4 shows a VHDL description of the adder.

```
package LOCAL is
  constant N:    INTEGER := 4;

  procedure BITSLICE(
      A, B: in BIT_VECTOR(3 downto 0);
      CIN: in BIT;
      signal S: out BIT_VECTOR(3 downto 0);
      signal GP, GG: out BIT);
  procedure PG(
      A, B: in BIT_VECTOR(3 downto 0);
      P, G: out BIT_VECTOR(3 downto 0));
  function SUM(A, B, C: BIT_VECTOR(3 downto 0))
      return BIT_VECTOR;
  procedure CLA(
      P, G: in BIT_VECTOR(3 downto 0);
      CIN: in BIT;
      C: out BIT_VECTOR(3 downto 0);
      signal GP, GG: out BIT);
end LOCAL;
```

Figure B-4

```
package body LOCAL is

   --------------------------------------------------
   -- Compute sum and group outputs from a, b, cin
   --------------------------------------------------

   procedure BITSLICE(
       A, B: in BIT_VECTOR(3 downto 0);
       CIN: in BIT;
       signal S: out BIT_VECTOR(3 downto 0);
       signal GP, GG: out BIT) is

     variable P, G, C: BIT_VECTOR(3 downto 0);
   begin
     PG(A, B, P, G);
     CLA(P, G, CIN, C, GP, GG);
     S <= SUM(A, B, C);
   end;


   ---------------------------------------------------
   -- Compute propagate and generate from input bits
   ---------------------------------------------------

   procedure PG(A, B: in BIT_VECTOR(3 downto 0);
                P, G: out BIT_VECTOR(3 downto 0)) is

   begin
     P := A or B;
     G := A and B;
   end;


   ---------------------------------------------------
   -- Compute sum from the input bits and the carries
   ---------------------------------------------------

   function SUM(A, B, C: BIT_VECTOR(3 downto 0))
       return BIT_VECTOR is

   begin
     return(A xor B xor C);
   end;
```

Continue Figure B-4

```
-------------------------------
-- 4-bit carry-lookahead block
-------------------------------

procedure CLA(
    P, G: in BIT_VECTOR(3 downto 0);
    CIN: in BIT;
    C: out BIT_VECTOR(3 downto 0);
    signal GP, GG: out BIT) is

  variable TEMP_GP, TEMP_GG, LAST_C: BIT;
begin
  TEMP_GP := P(0);
  TEMP_GG := G(0);
  LAST_C := CIN;
  C(0) := CIN;

  for I in 1 to N-1 loop
    TEMP_GP := TEMP_GP and P(I);
    TEMP_GG := (TEMP_GG and P(I)) or G(I);
    LAST_C := (LAST_C and P(I-1)) or G(I-1);
    C(I) := LAST_C;
  end loop;

  GP <= TEMP_GP;
  GG <= TEMP_GG;
  end;
end LOCAL;

use WORK.LOCAL.ALL;


-----------------------------------
-- A 32-bit carry-lookahead adder
-----------------------------------

entity ADDER is
  port(A, B: in BIT_VECTOR(31 downto 0);
       CIN: in BIT;
       S: out BIT_VECTOR(31 downto 0);
       COUT: out BIT);
end ADDER;
```

Continue Figure B-4

```
architecture BEHAVIOR of ADDER is

  signal GG,GP,GC: BIT_VECTOR(7 downto 0);
    -- First-level generate, propagate, carry
  signal GGG, GGP, GGC: BIT_VECTOR(3 downto 0);
    -- Second-level gen, prop, carry
  signal GGGG, GGGP: BIT;
    -- Third-level gen, prop

begin
  -- Compute Sum and 1st-level Generate and Propagate
  -- Use input data and the 1st-level Carries computed
  -- later.
  BITSLICE(A( 3 downto  0),B( 3 downto  0),GC(0),
           S( 3 downto  0),GP(0), GG(0));
  BITSLICE(A( 7 downto  4),B( 7 downto  4),GC(1),
           S( 7 downto  4),GP(1), GG(1));
  BITSLICE(A(11 downto  8),B(11 downto  8),GC(2),
           S(11 downto  8),GP(2), GG(2));
  BITSLICE(A(15 downto 12),B(15 downto 12),GC(3),
           S(15 downto 12),GP(3), GG(3));
  BITSLICE(A(19 downto 16),B(19 downto 16),GC(4),
           S(19 downto 16),GP(4), GG(4));
  BITSLICE(A(23 downto 20),B(23 downto 20),GC(5),
           S(23 downto 20),GP(5), GG(5));
  BITSLICE(A(27 downto 24),B(27 downto 24),GC(6),
           S(27 downto 24),GP(6), GG(6));
  BITSLICE(A(31 downto 28),B(31 downto 28),GC(7),
           S(31 downto 28),GP(7), GG(7));

  -- Compute first-level Carries and second-level
  -- generate and propagate.
  -- Use first-level Generate, Propagate, and
  -- second-level carry.
  process(GP, GG, GGC)
    variable TEMP: BIT_VECTOR(3 downto 0);
  begin
    CLA(GP(3 downto 0), GG(3 downto 0), GGC(0), TEMP,
        GGP(0), GGG(0));
    GC(3 downto 0) <= TEMP;
  end process;

  process(GP, GG, GGC)
    variable TEMP: BIT_VECTOR(3 downto 0);
  begin
    CLA(GP(7 downto 4), GG(7 downto 4), GGC(1), TEMP,
        GGP(1), GGG(1));
    GC(7 downto 4) <= TEMP;
  end process;
```

Continue Figure B-4

```
-- Compute second-level Carry and third-level
  --    Generate and Propagate
  -- Use second-level Generate, Propagate and Carry-in
  --    (CIN)
  process(GGP, GGG, CIN)
    variable TEMP: BIT_VECTOR(3 downto 0);
  begin
    CLA(GGP, GGG, CIN, TEMP, GGGP, GGGG);
    GGC <= TEMP;
  end process;

  -- Assign unused bits of second-level Generate and
  --    Propagate
  GGP(3 downto 2) <= "11";
  GGG(3 downto 2) <= "00";

  -- Compute Carry-out (COUT)
  -- Use third-level Generate and Propagate and
  --    Carry-in (CIN).
  COUT <= GGGG or (GGGP and CIN);
end BEHAVIOR;
```

Continue Figure B-4

In the carry-lookahead adder implementation, procedures perform the computation of the design. The procedures can also be written as separate entities and used by component instantiation, producing a hierarchical design. Note that the keyword `signal` is included before some of the interface parameter declarations, which is required for out formal parameters when the actual parameters must be signals.

The output parameter `c` from the `CLA` procedure is not declared as a signal, which makes it illegal to use in a concurrent procedure call (since only signals can be used in such calls). To overcome this problem, sub-processes are used, declaring a temporary variable `TEMP`. `TEMP` receives the value of the `c` parameter and assigns it to the appropriate signal.

C | VHDL Structure & Syntax

This appendix diagrams VHDL design hierarchy, structure, and language syntax. Its syntax is not official, but is simplified to aid usage and understanding. The syntax is based on the 1987 standard. For the formal VHDL syntax, refer to IEEE Standard VHDL Language Reference Manual. This chapter is organized into the following sections:

- Design Hierarchy
- Concurrent Statements
- Sequential Statements
- Specifications
- Use & Library Clauses
- Declarations
- Library Units
- Predefined Attributes
- Package STANDARD
- TEXTIO Package

C.1 Design Hierarchy

The figure below diagrams VHDL's design hierarchical organization.

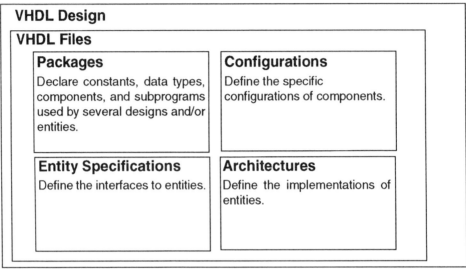

VHDL Design

VHDL Files

Packages
Declare constants, data types, components, and subprograms used by several designs and/or entities.

Configurations
Define the specific configurations of components.

Entity Specifications
Define the interfaces to entities.

Architectures
Define the implementations of entities.

Figure C-1

Architectures

The function of an entity is determined by its architecture. *Figure C-2* shows the organization of an architecture. Not all architectures contain every construct shown; the order of the concurrent statements in an architecture is not important.

Architecture

Declarations
Declare signals to communicate between concurrent statements and between concurrent statements and the interface ports. Declare types, constants, components, and subprograms used in the architecture.

Concurrent Statements

Blocks
Collect concurrent statements together.

Component Instantiations
Create an instance of another entity.

Signal Assignments
Compute values and assign them to signals.

Processes
Define a new algorithm.

Procedure Calls
Invoke a predefined algorithm.

Figure C-2

An architecture consists of a declaration section where *signals*, *types*, *constants*, *components*, and *subprograms* are declared, followed by a collection of concurrent statements.

Signals are used to connect the separate pieces of an architecture (the concurrent statements) to each other and to the outside world through interface ports. Each signal is declared to have a type that determines the kind of data it carries. Types, constants, components, and subprograms declared in an architecture are local to that architecture. To use these declarations in more than one entity or architetcure, place them in a package.

Each concurrent statement in an architecture defines a unit of computation that reads signals, performs a computation based on their value, and assigns computed values to signals. Concurrent statements compute their values all at the same time. While the order of concurrent statements has no effect on execution order, concurrent statements often coordinate processing by communicating with each other via signals. There are five kinds of concurrent statements:

- Blocks - collect the concurrent statements.

- Signal assignments - assign computed values to signals or interface ports.

- Procedure calls- call algorithms that compute and assign values to signals.

- Component instantiations - create an instance of an entity, connecting its interface ports to signals and/or interface ports of the entity being defined.

- Processes - define sequential algorithms which read the values of signals, and compute new values to assign to other signals.

Processes

Processes contain sequential statements that define algorithms. Unlike concurrent statements, sequential statements execute one after another, which allows you to perform step-by-step computations. Processes, like all concurrent statements, read and write signals and interface port values to communicate with the rest of the architecture and with the enclosing system.

Process

Declarations

Internal variables that hold temporary values in the sequence of computations, as well as types, constants, components, and subprograms used locally.

Sequential Statements

Signal assignments Compute values and assign them to signals.	**LOOP Statements** Execute statements repeatedly.
Procedure Calls Invoke predefined algorithms.	**NEXT statements** Skip remainder of LOOP.
Variable statements Store partial results in variables.	**EXIT statements** Terminate the execution of a LOOP.
IF statements Conditionally execute groups of sequential statements.	**WAIT statements** Wait for a clock signal.
CASE statements Select a group of sequential statements to execute.	**NULL statements** Place holders that perform no action.

Figure C-3

Processes are unique in that they behave like concurrent statements to the rest of the design, but they are internally sequential. In addition, only processes can define variables, which hold intermediate values in a sequence of computations (VHDL 87).

Since the statements in a process are sequentially executed, there are several constructs that control the order in which the statements are executed.

Subprograms

Subprograms, like processes, use sequential statements to define algorithms that compute values. Unlike processes, they may not directly read or write signals from the rest of the architecture. All communication is through the subprogram's interface. Each subprogram call has its own set of interface signals.

The construct organization of a subprogram is the same as that of processes, with one added sequential statement, the RETURN statement. A RETURN statement terminates the execution of a subprogam.

The two types of subprograms are functions and procedures: a function returns a value directly, while a procedure returns zero or more values through its interface. Subprograms are useful because they can be reused to perform repeated calculations.

Packages

Packages collect constants, data types, component declarations, and subprograms that will be used by several designs or entities. *Figure C-4* shows the typical organization of a package. Some packages may not use all of the constructs listed.

Packages

Constants
Declare constant values used by a design.

Types
Declare the data types used by a design.

Component Declarations
Declare interfaces for design entities.

Subprograms
Declare algorithms used by the design.

Figure C-4

The elements contained in the package are described below:

- Constants - in packages declare system-wide parameters, such as data-path widths.

- VHDL data-type declarations - are included in a package to define the types of data used in the design. Especially in interface declarations, it is important that all entities in a design use common types (for example, a design's internal address type).

- Component declarations - specify the interfaces to entities that can be instantiated in the design.

- Subprograms - define algorithms that can be called anywhere in a design.

Packages can be made general to be used in many different designs. For example, the ARITHMETIC package defines the data types signed and unsigned and provides useful functions that operate on them.

C.2 Concurrent Statements

Concurrent statements execute at the same time. They are:

- Block Statement
- Component Instantiation Statement
- Concurrent Assertion Statement
- Concurrent Procedure Call
- Concurrent Signal Assignment Statement
- Generate Statement
- Process Statement

Block Statement

Concurrent Statements

block_statement

```
label: block [(guard_expression)]
    [generic]
    [ports]
    [declarations]
begin
concurrent_statements
end block [label];
```

Block

```
A: block

begin..............

block_statement

end block A;
```

Architecture

```
architecture A of E is



begin...............

block_statement

end A;
```

Concurrent Statements   Component Instantiation Statement

---

component_instantiation_statement

```
label : name [generic map (map)] [port map (signals)];
```

Block

```
A: block

begin...................

component_instantiation_statement

end block A;
```

Architecture

```
architecture A of E is

begin...............

component_instantiation_statement

end A;
```

Concurrent Assertion Statement   Concurrent Statements

---

concurrent_assertion_statement

```
assert condition
      [report string_expression]
      [severity {NOTE|WARNING|ERROR|FAILURE}];
```

Entity

```
entity E is

begin.............

concurrent_assertion_statement*

end E;
```

Block

```
A: block

begin....................

concurrent_assertion_statement

end block A;
```

Architecture

```
architecture A of E is

begin................

concurrent_assertion_statement

end A;
```

*Typical use for error checking.

Concurrent Statements   Concurrent Procedure Call

concurrent_procedure_call

```
[label:] procedure_name [(parameters)];
```

Entity

```
entity E is

begin................

concurrent_procedure_call

end E;
```

Block

```
A: block

begin................

concurrent_procedure_call

end block A;
```

Architecture

```
architecture A of E is

begin...............

concurrent_procedure_call

end A;
```

Concurrent Signal Assignment Statement

Concurrent Statements

concurrent_signal_assignment_statement

```
[label:][{conditional_assignment|assignment|selected_assignment}];
```

Block

```
A: block

begin....................

concurrent_signal_assignment_statement

end block A;
```

Architecture

```
architecture A of E is

begin................

concurrent_signal_assignment_statement

end A;
```

Concurrent Statements

Generate Statement

---

generate_statement

```
label :[{for specification | if condition}] generate
      concurrent_statements
end generate [label];
```

Block

```
A: block

begin....................

generate_statement

end block A;
```

Architecture

```
architecture A of E is

begin................

generate_statement

end A;
```

Process Statement   ## Concurrent Statements

---

process_statement

```
[label:] process [(sensitivity_list)]
[variable_declaration]
[type_declaration]
[subprogram_declaration]
[declarations]
begin
sequential_statements*
end process [label];
```

Entity

```
entity E is

begin..............

process_statement

end E;
```

Block

```
A: block

begin..............

process_statement

end block A;
```

Architecture

```
architecture A of E is

begin................

process_statement

end A;
```

* Cannot contain a **wait** statement if sensitivity_list is used.

C.3 Sequential Statements

Sequential statements define algorithms for the execution of a subprogram or process. They execute in order in which they appear. They are:

- Assertion Statement
- Case Statement
- Exit Statement
- If Statement
- Loop Statement
- Next Statement
- Null Statement
- Procedure Call Statement
- Return Statement
- Signal Assignment Statement
- Variable Assignment Statement
- Wait Statement

Assertion Statement

---

assertion_statement

assert condition*
    [**report** string_expression]
    [**severity** {**NOTE**|**WARNING**|**ERROR**|**FAILURE**}];

Process

```
A: process


begin..............

assertion_statement

end process A;
```

Procedure

```
procedure A (...) is


begin..............

assertion_statement

end A;
```

*When condition is <u>false,</u> [string_expression] is printed.

Sequential Statements

Case Statement

---

case_statement

```
case expression* is
when choices-1 => sequential_statements

    .

    .

when choices-n => sequential_statements

end case;
```

Process

```
A: process



begin.................

case_statement

end process A;
```

Procedure

```
procedure A (...) is



begin....................

case_statement

end A;
```

*Avoid the use of parenthesis, if possible.

Exit Statement

Sequential Statements

exit_statement

exit [label] [**when** condition];

Process

```
A: process


begin....................
loop....................
exit_statement

end process A;
```

Procedure

```
procedure A (...) is


begin....................
loop....................
exit_statement

end A;
```

---

if_statement

```
if condition then
   sequential_statements
   {elsif condition then sequential_statements}
   [else sequential_statements]
end if;
```

Process

```
A: process


begin...................

if_statement

end process A;
```

Procedure

```
procedure A (...) is


begin....................

if_statement

end A;
```

Loop Statement

Sequential Statements

loop_statement

```
[label:] [while condition |for loop_specification] loop
    sequential_statements
end loop [label];
```

Process

```
A: process


begin...................

loop_statement

end process A;
```

Procedure

```
procedure A (...) is


begin...................

loop_statement

end A;
```

Sequential Statements   Next Statement

---

next_statement

```
next [label] [when condition];
```

Process

```
A: process


begin...................
loop...................
next_statement

end process A;
```

Procedure

```
procedure A (...) is


begin...................
loop...................
next_statement

end A;
```

Null Statement

null_statement

```
null;
```

Process

```
A: process


begin...................

null_statement

end process A;
```

Procedure

```
procedure A (...) is


begin...................

null_statement

end A;
```

Sequential Statements

Procedure Call Statement

procedure_call_statement

```
procedure_name [(parameters)] ;
```

Process

```
A: process


begin...................

procedure_call_statement

end process A;
```

Procedure

```
procedure A (...) is


begin....................

procedure_call_statement

end A;
```

Return Statement

Sequential Statements

---

return_statement–F

```
return expression;
```

return_statement–P

```
return;
```

Function

```
function A (...) is


begin...................

return_statement-F

end A;*
```

Procedure

```
procedure A (...) is


begin...................

return_statement-P

end A;
```

*You must execute a **return** statement.

Sequential Statements

Signal Assignment Statement

signal_assignment_statement

```
target <= expression [after time_expression]
           .
           .
    {,expression [after time_expression} ;
```

Process

```
A: process


begin...................

signal_assignment_statement

end process A;
```

Procedure

```
procedure A (...) is


begin...................

signal_assignment_statement

end A;
```

Variable Assignment Statement

Sequential Statements

variable_assignment_statement

```
target := expression;
```

Process

```
A: process


begin....................

variable_assignment_statement

end process A;
```

Procedure

```
procedure A (...) is


begin....................

variable_assignment_statement

end A;
```

Sequential Statements   Wait Statement

---

wait_statement

```
wait
    [on signal_name {,signal_name}]
    [until conditional_expression]
    [for time_expression];
```

Process

```
A: process


begin..............

wait_statement

end process A;
```

Procedure

```
procedure A (...) is


begin............

wait_statement*

end A;
```

*A **Function** may not contain a **wait_statement**.

C.4 Specifications

Specifications associate additional information with a VHDL description. A specification associates additional information with a previously pndeclared entity. There are three kinds of specifications:

- Attribute Specifications
- Configuration Specifications

Specifications   Attribute Specification

---

attribute_specification

> **attribute** attribute_name
>   **of** entity_name **is** expression;

Entity

```
entity E is
attribute_specification

begin..............

end E;
```

Process

```
A: process
attribute_specification

begin..............

end process A;
```

Package Declaration

```
package A is
attribute_specification

end A;
```

Block

```
A: block
attribute_specification

begin..............

end block A;
```

Procedure

```
procedure A (...) is
attribute_specification

begin..............

end A;
```

Architecture

```
architecture A of E is
attribute_specification

begin................

end A;
```

Configuration

```
configuration A of E is
attribute_specification

end A;
```

Configuration Specification

---

configuration_specification

```
for component_name
use [generic_map_part]
      [port_map_part];
```

Block

```
A: block
configuration_specification

begin...................

end block A
```

Architecture

```
architecture A of E is
configuration_specification

begin...............

end A;
```

C.5 Library & USE Clauses

Clauses in VHDL select and define declarations. A LIBRARY clause defines logical names for design libraries in the host environment. The USE clause selects declarations made visible by the selection.

LIBRARY Clause

Clauses

---

library_clause

```
LIBRARY names;
```

Entity

```
library_clause
entity E is

begin..............

end E;
```

Package Declaration

```
library_clause
package A is

end A;
```

Package Body

```
library_clause
package body A is

end A;
```

Architecture

```
library_clause
architecture A of E is

begin................

end A;
```

Configuration

```
library_clause
configuration A of E is

end A;
```

Clauses   USE Clause

---

use_clause

```
USE selected_names;
```

Entity

```
use_clause
entity E is
use_clause

begin..............

end E;
```

Process

```
A: process
use_clause

begin..............

end process A;
```

Package Declaration

```
use_clause
package A is
use_clause

end A;
```

Block

```
A: block
use_clause

begin..............

end block A;
```

Procedure

```
procedure A (...) is
use_clause

begin..............

end A;
```

Package Body

```
use_clause
package body A is
use_clause
```

Architecture

```
use_clause
architecture A of E is
use_clause

begin................

end A;
```

Configuration

```
use_clause
configuration A of E is
use_clause

end A;
```

C.6 Declarations

VHDL defines several kinds of entities that are declared explicitly or implicitly by declarations. The VHDL declarations are:

- Alias Declaration
- Attribute Declaration
- Component Declaration
- Constant Declaration
- File Declaration
- Signal Declaration
- Subprogram Declaration
- Subprogram Body
- Subtype Declaration
- Type Declaration
- Variable Declaration

Declarations

Alias Declaration

alias_declaration

```
alias name1 : type [(indexes)] is name2 [(indexes)];
```

Entity

```
entity E is
alias_declaration

begin..............

end E;
```

Process

```
A: process
alias_declaration

begin..............

end process A;
```

Package Declaration

```
package A is
alias_declaration

end A;
```

Block

```
A: block
alias_declaration

begin..............

end block A;
```

Procedure

```
procedure A (...) is
alias_declaration

begin..............

end A;
```

Package Body

```
package body A is
alias_declaration

end A;
```

Architecture

```
architecture A of E is
alias_declaration

begin................

end A;
```

Attribute Declaration   **Declarations**

---

attribute_declaration

```
attribute name: type;
```

Entity

```
entity E is
```
attribute_declaration

```
begin..............

end E;
```

Process

```
A: process
```
attribute_declaration

```
begin..............

end process A;
```

Package Declaration

```
package A is
```
attribute_declaration

```
end A;
```

Block

```
A: block
```
attribute_declaration

```
begin..............

end block A;
```

Procedure

```
procedure A (...) is
```
attribute_declaration

```
begin..............

end A;
```

Architecture

```
architecture A of E is
```
attribute_declaration

```
begin...............

end A;
```

Declarations   Component Declaration

---

component_declaration

```
component identifier :
   [generic (generic_list);]
   [port (port_list);]

end component;
```

Package Declaration

```
package A is
component_declaration


end A;
```

Block

```
A: block
component_declaration

begin..............

end block A;
```

Architecture

```
architecture A of E is
component_declaration

begin................

end A;
```

Constant Declaration

Declarations

---

constant_declaration

    constant name :type := expression;

    constant name :array_type [(indexes)] := expression;

Entity
```
entity E is
constant_declaration

begin..............


end E;
```

Process
```
A: process
constant_declaration

begin..............

end process A;
```

Package Declaration
```
package A is
constant_declaration*


end A;
```

Block
```
A: block
constant_declaration

begin..............

end block A
```

Procedure
```
procedure A (...) is
constant_declaration

begin..............

end A;
```

Package Body
```
package body A is
constant_declaration


end A;
```

Architecture
```
architecture A of E is
constant_declaration
begin...............

end A;
```

* expression is optional in the *Package Declaration* if it is a deferred constant.

Declarations

File Declaration

---

file_declaration

```
file name:type is [mode] logical_name;
```

Entity

```
entity E is
file_declaration

begin.............

end E;
```

Process

```
A: process
file_declaration

begin.............

end process A;
```

Package Declaration

```
package A is
file_declaration

end A;
```

Block

```
A: block
file_declaration

begin.............

end block A;
```

Procedure

```
procedure A (...) is
file_declaration

begin.............

end A;
```

Package Body

```
package body A is
file_declaration

end A;
```

Architecture

```
architecture A of E is
file_declaration
begin..............

end A;
```

Signal Declarations

Declarations

---

signal_declaration

```
signal names :type [constraint] [:= expression];
        or, in an entity
```

port_declaration

```
port (names :direction type [:= expression] [;more_signals]);
```

Entity

```
entity E is
port_declaration
signal_declaration
begin..............

end E;
```

Package Declaration

```
package A is
signal_declaration


end A;
```

Block

```
A: block
signal_declaration

begin..............

end block A;
```

Architecture

```
architecture A of E is
signal_declaration

begin...............

end A;
```

Declarations

Subprogram Declaration
Subprogram Body

---

subprogram_declaration

```
procedure name [(parameters)]
| function name [(parameters)]

  return type
  ;
```

subprogram_body

```
is
   declarations
begin
   sequential_statements
end [name];
```

Entity

```
entity E is
subprogram_declaration
subprogram_body
begin..............


end E;
```

Process

```
A: process
subprogram_declaration
subprogram_body
begin..............

end process A;
```

Package Declaration

```
package A is
subprogram_declaration;

end A;
```

Block

```
A: block
subprogram_declaration
subprogram_body
begin..............

end block A;
```

Procedure

```
procedure A (...) is
subprogram_declaration
subprogram_body
begin..............

end A;;
```

Package Body

```
package body A is
subprogram_declaration
subprogram_body


end A;
```

Architecture

```
architecture A of E is
subprogram_declaration
subprogram_body
begin..............
end A;
```

Subtype Declaration

Declarations

subtype_declaration

subtype name **is** [resolution_function] type[constraint];

Entity

```
entity E is
subtype_declaration

begin..............

end E;
```

Process

```
A: process
subtype_declaration

begin..............

end process A;
```

Package Declaration

```
package A is
subtype_declaration

end A;
```

Block

```
A: block
subtype_declaration

begin..............

end block A;
```

Procedure

```
procedure A (...) is
subtype_declaration

begin..............

end A;
```

Package Body

```
package body A is
subtype_declaration

end A;
```

Architecture

```
architecture A of E is
subtype_declaration

begin................

end A;
```

Declarations   Type Declaration

---

type_declaration

```
type name is definition;
type name; *
```

Entity

```
entity E is
type_declaration

begin.............

end E;
```

Process

```
A: process
type_declaration

begin.............

end process A;
```

Package Declaration

```
package A is
type_declaration

end A;
```

Block

```
A: block
type_declaration

begin.............

end block A;
```

Procedure

```
procedure A (...) is
type_declaration

begin.............

end A;
```

Package Body

```
package body A is
type_declaration

end A;
```

Architecture

```
architecture A of E is
type_declaration

begin................

end A;
```

*Incomplete declaration, which requires the syntax above it to follow.

Variable Declaration

Declarations

---

variable_declaration

```
variable names : type [constraint] [:= expression];
```

Process

```
A: process
variable_declaration

begin..............

end process A;
```

Procedure

```
procedure A (...) is
variable_declaration

begin..............

end A;
```

C.7 Library Units

Library units are the basic building blocks of VHDL. They represent the main components of the language, which represent a design. The VHDL library units are:

- Architecture Body
- Configuration Declaration
- Entity Declaration
- Package Body
- Package Declaration

Architecture Body

Library Units

---

architecture_body

```
architecture name of
entity_name is
   [types]
   [constants]
   [signals]
   [subprograms]
   [other declarations]
begin
   concurrent_statements
end [name];
```

Architecture

```
architecture A of E is

begin................

end A;
```

Library Units

Configuration Declaration

---

configuration_declaration

```
configuration name of
entity_name is
    declarative_part
    block_configuration
end [name];
```

Configuration

```
configuration A of E is



end A;
```

Entity Declaration

Library Units

---

entity_declaration

```
entity name is
    [generics][ports]
    [declarations]
[ begin statements]*
end [name];
```

Entity

```
entity E is


begin..............


end E;
```

---

* Typically, an entity does not have statements. If it does, the statements cannot operate on signals.

---

package_body

```
package_body name is
    [subprogram]
    [type]
    [constant]
    [signal]
    [declarations]
    end [name];
```

Package Body

```
package body A is



end A;
```

* Optional Body

Package Declaration

Library Units

---

package_declaration

```
package name is
   [subprogram]
   [type]
   [constant]
   [signal]
   [file]
   [alias]
   [USE clause]
   [declarations]
   end [name];
```

Package Declaration

```
package A is



end A;


```

C.8 Predefined Attributes

Predefined attributes denote values, functions, types, and ranges associated with different kinds of entities. There are four kinds of attributes:

- •Array-related attributes
- •Signal attributes
- •Type-related attributes

Also listed are attributes new to VHDL '92

Array-Related Attributes

Attribute Name	Prefix	Parameter	Result
A'LEFT[(N)]	For an array or type object.	Integer expression	Left bound of the Nth index range of A.
A'RIGHT[(N)]	For an array or type object.	Integer expression	Right bound of the Nth index range of A.
A'HIGH[(N)]	For an array or type object.	Integer expression	Upper bound of the Nth index range of A.
A'LOW[(N)]	For an array or type object.	Integer expression	Lower bound of the Nth index range of A.

(Continue Array-Related Attributes)

Attribute Name	Prefix	Parameter	Result
A'LENGTH[(N)]	Array or type object.	Integer expression where the value must not exceed the dimensions of A. Default is 1.	Number of values in the Nth index range.
A'RANGE[(N)]	A prefix A for an array object, an alias, or a constrained array subtype.	Integer expression	The range A'LEFT(N) to A'RIGHT(N) A'LEFT(N) downto A'RIGHT(N) of A is descending.
A' REVERSE_ RANGE[(N)]	An array or type object	Integer expression	The range A'RIGHT(N) to A'LEFT(N). The range A'RIGHT(N) downto A'LEFT(N) if the Nth index range of A is descending.

Signal Attributes

Attribute Name	Prefix	Parameter	Result
S'DELAYED(T)	Signal S.	Expression of type TIME that evaluates to a non-negative value.Default 0ns.	A signal equivalent to S delayed T units of time.
S'QUIET(T)	Signal S.	Expression of type TIME that evaluates to a non-negative value.Default is 0ns.	A BOOLEAN signal where value = TRUE when the signal has been quiet t units.
s'STABLE (T)	Signal S.	Expression of type TIME that evaluates to a non-negative value.Default is 0ns.	A BOOLEAN signal which is true whenever the reference signal has had no events for time T
S'TRANSACTION	Signal S.		A bit signal with a value that toggles in each simulation cycle in which S is active.
S'EVENT	Signal S.		A boolean value that indicates if an event has occurred on S.
S'ACTIVE	Signal S.		A boolean value that indicates if S is active.
S'LAST_EVENT	Signal S.		Amount of elapsed time since last event on signal S.
S'LAST_ACTIVE	Signal S.		Amount of elapsed time since S was active.
S'LAST_VALUE	Signal S.		Previous value of S before last change.

Type-Related Attributes

Attribute Name	Prefix	Parameter	Result
T'BASE	Any type T.		Base type of T.
T'LEFT	Any scalar type T.		Left bound of type T.
T'RIGHT	Any scalar type T.		Right bound of type T.
T'HIGH	Any scalar type T.		Upper bound of type T.
T'LOW	Any scalar type or subtype T.		Lower bound of type T.
T'POS(X)	Any discrete or physical type T.	An expression whose type is the base of type T.	Position number of the value of X in type T.
T'VAL(X)	Any discrete or physical type T.	An expression of any integer type.	Value whose value corresponds to position X in T.
T'SUCC(X)	Any discrete or physical type T.	An expression whose type is the base type of T.	Value whose position number is one greater than that of X.
T'PRED(X)	Any discrete or physical type T.	An expression whose type is the base type of T.	Value whose position number is one less than that of X.
T'LEFTOF(X)	Any discrete or physical type T.	An expression whose type is the base type of T.	Value to the left of the X in T.
T'RIGHTOF(X)	Any discrete or physical type T.	An expression whose type is the base type of T.	Value to the right of the X in T.

New Attributes in VHDL 92

```
T'ascending

T'image

T'value

S'driving

S'driving_values

E'path_name

E'simple_name
```

C.9 Package STANDARD

```
--predefined enumeration types:

type Boolean is (FALSE, TRUE);
type BIT is ('0','1');

type CHARACTER is (
NUL, SOH, STX, ETX, EOT, ENQ,ACK, BEL, BS, HT, LF, VT, FF, CR, SO,
SI, DLE, DC1, DC2, DC3, DC4, NAK, SYN, ETB, CAN, EM, SUB, ESC, FSP,
GSP, RSP, USP,
' ', '!', '"', '#', '$', '%', '&', ''', '(', ')', '*', '+', ',',
'-','.', '/','0', '1', '2', '3', '4', '5', '6', '7', '8', '9', ':',
';', '<', '=', '>', '?','@', 'A', 'B', 'C', 'D', 'E', 'F', 'G', 'H',
'I', 'J', 'K', 'L', 'M', 'N', 'O','P', 'Q', 'R', 'S', 'T', 'U', 'V',
'W', 'X', 'Y', 'Z', '[', '\', ']', '^', '_' '`', 'a', 'b', 'c',
'd',, 'e', 'f', 'g', 'h', 'i', 'j', 'k', 'l', 'm' 'n', 'o',
'p', 'q', 'r', 's', 't', 'u', 'v',, 'w', 'x', 'y', 'z','{', |,'}',
'~', DEL);

TYPE SEVERITY_LEVEL IS (NOTE, WARNING, ERROR, FAILURE);

--predefined numeric types:

type INTEGER is range implementation_defined;
type REAL is range implementation_defined;

--predefined type TIME:
type TIME is range implementation_defined
     units
         fs;--femtosecond
         ps=1000fs;--picosecond
         ns=1000ps;--nanosecond
         us=1000ns--microsecond
         ms=1000us--millisecond
         sec=1000ms;--second
         min=60 sec--minute
         hr=60 min;--hour
   end units;

--function that returns current simulation time:
function NOW return TIME;

--predefined numeric subtypes:
subtype NATURAL is INTEGER range 0 to INTEGER'HIGH;
subtype POSITIVE is INTEGER range 1 to INTEGER'HIGH;

--predefined array types:
type STRING is array (POSITIVE range <>) of CHARACTER;
type BIT_VECTOR is array (NATURAL range <>) of BIT;
end STANDARD;
```

C.10 TEXTIO Package

```
Package TEXTIO is

--Type Definitions for Text I/O
type LINE is acces STRING;---a LINE is a pointer to a string value
type TEXT is file of STRING;-a file of variable-length ASCII
                            records
type SIDE is (RIGHT, LEFT);--for justifying output data within
                           fields
subtype WIDTH is NATURAL;----for specifying widths of output fields

--Standard Text Files
file INPUT: TEXT is in "STD_INPUT";
file OUTPUT: TEXT is out "STD_OUTPUT"

--Input Routines for Standard Types
procedure READLINE (F: in TEXT; L: out LINE);

procedure READ(L:inout LINE;VALUE:out BIT_VECTOR;GOOD:out BOOLEAN);
procedure READ (L:inout LINE; VALUE: out BIT_VECTOR);

 --OUTPUT ROUTINES FOR STANDARD TYPES
PROCEDURE WRITELINE (F: OUT TEXT; L: IN LINE);
PROCEDURE WRITE (L: inout LINE; VALUE: IN BIT;
        JUSTIFIED: in SIDE := RIGHT; FIELD: in WIDTH :=0;
PROCEDURE WRITE (L: inout LINE; VALUE: IN BIT_VECTOR;
        JUSTIFIED: in SIDE := RIGHT; FIELD: in WIDTH :=0;
PROCEDURE WRITE (L: inout LINE; VALUE: IN BOOLEAN;
        JUSTIFIED: in SIDE := RIGHT; FIELD: in WIDTH :=0;
PROCEDURE WRITE (L: inout LINE; VALUE: IN CHARACTER;
        JUSTIFIED: in SIDE := RIGHT; FIELD: in WIDTH :=0;
PROCEDURE WRITE (L: inout LINE; VALUE: IN INTEGER;
        JUSTIFIED: in SIDE := RIGHT; FIELD: in WIDTH :=0;
PROCEDURE WRITE (L: inout LINE; VALUE: in REAL;
        JUSTIFIED in SIDE:= RIGHT; FIELD: in WIDTH :=0;
        DIGITS: in NATURAL :=0);
PROCEDURE WRITE (L: inout LINE; VALUE: IN STRING;
        JUSTIFIED: in SIDE := RIGHT; FIELD: in WIDTH :=0;
PROCEDURE WRITE (L:inout LINE; VALUE: in TIME;
        JUSTIFIED: in SIDE := RIGHT; FIELD: in WIDTH:= 0;
        UNIT: in TIME := ns);

--File Position Predicates
function ENDLINE (L: in LINE)return BOOLEAN;
--function ENDFILE (F: in TEXT) return BOOLEAN;

end TEXTIO;
```

INDEX

Symbols

Numerics

A

B

C

D

E

M

N

O

P

Bibliography

D. Coelho, *The VHDL Handbook*, Kluwer Academic Publishers, 1989, Norwell, MA.

T.E. Dillinger, et al, *A Logic Synthesis System for VHDL Design Description*, IEEE ICCAD-89, Santa Clara, CA.

R. Lipsett, C. Shaefer, C. Ussery, *VHDL Hardware Description and Design*, Intermetrics, 1989.

D. Perry, *VHDL*, McGraw Hill, 1989, NY.

S. Leung, M. Shanblatt, *ASIC System Design with VHDL*, A Paradigm, Kluwer Academic Publishers, 1989, Norwell, MA.

IEEE, *IEEE Standard VHDL Language Reference Manual*, 1988, NY.

R.E. Harr, et al, *Applications of VHDL to Circuit Design*, Kluwer Academic Publishers, 1991, Norwell, MA.

Jean-Michel Berge, Alain Fonkoua, Serge Maginot, Jacques Rouillard, *VHDL's Designer's Reference*, Kluwer Academic Publishers, 1992, Norwell, MA.

Jayaram Bhasker, *A VHDL Primer*, Prentice-Hall, 1992, Englewood Cliffs, New Jersey.

Steve Carlson, *Introduction to HDL-Based Design Using VHDL*, 1991, Synopsys, Mt. View, CA.

Doug Dunlop, Cary Ussery, *An Introduction to VHDL 92*, VHDL International, VIUF Spring Conference, April 1992